Current Topics in Neurotoxicity

Series Editors
Richard M. Kostrzewa East Tennessee State University, Kostrzewa, USA
Trevor Archer Department of Psychology, Göteborg University, Göteborg, Sweden

For further volumes:
http://www.springer.com/series/8791

Keith A. Foster

Editor

Clinical Applications of Botulinum Neurotoxin

 Springer

Editor
Keith A. Foster
Syntaxin Ltd
Units 4–10 The Quadrant
Barton Lane
Abingdon
OX14 3YS
Oxfordshire
United Kingdom

ISBN 978-1-4939-0260-6   ISBN 978-1-4939-0261-3 (eBook)
DOI 10.1007/978-1-4939-0261-3
Springer New York Dordrecht Heidelberg London

Library of Congress Control Number: 2014930101

Printed on acid-free paper

Springer is part of Springer Science+Business Media (www.springer.com)

Contents

Contributors

Alberto Albanese Istituto di Neurologia, Università Cattolica del Sacro Cuore, Fondazione IRCCS Istituto Neurologico Carlo Besta, Milan, Italy

Anna Rita Bentivoglio Istituto di Neurologia, Università Cattolica del Sacro Cuore, Fondazione IRCCS Istituto Neurologico Carlo Besta, Milan, Italy

John Chaddock Syntaxin Ltd., Abingdon, UK

Roger R. Dmochowski Department of Urologic Surgery, Vanderbilt University Medical Center, Nashville, TN, USA

Dirk Dressler Movement Disorders Section, Department of Neurology, Hannover Medical School, Hannover, Germany

Keith A. Foster Syntaxin Ltd., Abingdon Oxfordshire, UK

Alex Gomelsky Department of Urology, Louisiana State University Health—Shreveport, Shreveport, LA, USA

Arianna Guidubaldi Istituto di Neurologia, Università Cattolica del Sacro Cuore, Fondazione IRCCS Istituto Neurologico Carlo Besta, Milan, Italy

Bahman Jabbari Department of Neurology, Yale University School of Medicine, New Haven, CT, USA

Duarte G. Machado Department of Neurology, Yale University School of Medicine, New Haven, CT, USA

Andy Pickett Toxin Science Limited, Wrexham, UK

Botulinum Research Center, UMASS Dartmouth, North Dartmouth, MA, USA

List of Abbreviations

AAN	American Academy of Neurology
ACh	Acetylcholine
ADLs	Activities of daily living
API	Active pharmaceutical ingredient
ATP	Adenosine-5'-triphosphate
ATP	Adenosine triphosphate
AUR	Acute urinary retention
BC	Bladder capacity
BCG	Bacillus calmette-guerin
BoNT	Botulinum neurotoxin
BoNT/A	BoNT type A
BPH	Benign prostate hyperplasia
BRV	Bladder reflex volume
BT	Botulinum toxin
CAMR	Centre for Applied Microbiology and Research
CD	Cervical dystonia
CDC	Centers for Disease Control & Prevention
CDH	Chronic daily headache
CFR	Code of federal regulations
cGMP	Current good manufacturing practice
CGRP	Calcitonin gene-related peptide
CIC	Clean intermittent catheterization
CLE	Chronic lateral epicondylitis
CM	Chronic migraine
CMG	Cystometography
CP	Cerebral palsy
DBS	Deep brain stimulation
DP	Drug product
DS	Drug substance
DSD	Detrusor sphincter dyssynergia
DVIU	Direct visual internal urethrotomy
EGF	Epidermal growth factor

EM	Episodic migraine
EMDA	Electromotive Drug Administration
EMEA	European Medicines Evaluation Agency
EMG	Electromyography
ET	Essential tremor
ETS	Endoscopic thoracic sympathectomy
FBC	Functional bladder capacity
FDA	Food and Drug Administration
GAP43	Growth associated protein 43
GFAP	Glial fibrillary acidic protein
HAS	Human serum albumin
HC	Heavy Chain
HD	Hydrodistention
HEPA	High efficiency particulate arrestant
HFS	Hemifacial spasm
HHS	United States Health and Human Services
IC	Interstitial cystitis
IDO	Idiopathic detrusor overactivity
IPSS	International prostate symptom score
LC	Light chain
LES	Lower oesophageal sphincter
LIBP	Lanzhou Institute for Biological Products
LUTS	Lower urinary tract symptoms
MBC	Mean bladder capacity
MBRV	Maximum bladder reflex volume
MCC	Maximum cystometric capacity
MDP	Maximum detrusor pressure
MFPS	Myofascial pain syndrome
MS	Multiple sclerosis
MSA	Multiple system atrophy
MU	Mouse units
NAPS	Neurotoxin-associated proteins
NDO	Neurogenic detrusor overactivity
NGF	Nerve growth factor
OAB	Overactive bladder
OCABR	Official control authority batch release
OMCL	Official medicines control laboratory
OMD	Oromandibular dystonia
PBSV	Painful bladder syndrome
PD	Parkinson's disease
PFV	Plantar faciitis
PGAC	Physicians global assessment of change
PHN	Post herpetic neuralgia
PREEMPT	Phase 3 research evaluating migraine prophylaxis therapy
PS	Piriformis syndrome

PSA	Prostate-specific antigen
PTN	Post-traumatic neuralgia
PVR	Post void residual
QC	Quality control
QOL	Quality of life
RCT	Randomised controlled trials
RMF	Rat muscle force
SCI	Spinal cord injury
SD	Spasmodic dystonia
SNAP	Synaptosomal associated membrane protein
SNAP25	Synaptosome-associated protein 25 kDa
SNARE	Soluble N-ethylmaleimide-sensitive factor attachment protein receptor
SV2	Synaptic vesicle glycoprotein 2
SVs	Synaptic vesicles
TD	Tardive dyskinesias
TeNT	Tetanus toxin
TH	Tension headache
TKA	Total knee arthroplasty
tPts	Trigger points
TPV	Total prostate volume
TRUS	Transrectal ultrasound
TSI	Targeted secretion inhibitor
TVEMP	Targeted vesicular exocytosis modulating protein
TWSTRS	Toronto western spasmotic torticollis rating scale
UES	Upper oesophageal sphincter
UMN	Upper motor neuron
UTI	Urinary tract infection
VAS	Visual analog scale
VAS	Visual analog score
VUR	Vesicoureteral reflux
WC	Writer's cramp
WDR	Wide dynamic range
WGA	Wheat germ agglutinin

Chapter 1
Overview and History of Botulinum Neurotoxin Clinical Exploitation

Keith A. Foster

Abstract Botulinum neurotoxin is a highly successful therapeutic agent used for the treatment of a range of severe, chronic diseases, and is also widely used and recognised as a cosmetic agent for reduction of facial wrinkles. And yet, this blockbuster therapeutic product is also the most lethal toxin known and a Centers for Disease Control and Prevention (CDC) Category A bioweapons threat. These apparently conflicting applications of the same agent have their origins in the unique biological properties of this fascinating protein. Unravelling the biology of the neurotoxin has informed understanding of the basis of both its toxicity and therapeutic activity. The unique properties of the neurotoxin have led to its becoming a significant therapeutic agent of benefit in an ever-expanding range of diseases of neuronal hyperactivity. Establishing the structural basis of its activity has opened up opportunities to engineer the toxin to create novel proteins of increased therapeutic effect and potential.

Keywords Botulinum neurotoxin · Therapeutic agent · Botulism · Neurotoxin-associated proteins · Neurotoxin complex · Oculinum · Botox® · Dysport® · Xeomin® · Myobloc®/NeuroBloc® · Recombinant protein

1.1 Introduction

Botulinum neurotoxin (BoNT), of which there are seven serotypes A–G (BoNT/A–/G), holds a unique status in public perception. It is the most lethal acute toxin known, with an estimated human lethal dose of 1.3–2.1 ng/kg intravenously or intramuscularly and 10–13 ng/kg when inhaled [1], and at the same time, it is a highly successful therapeutic agent that is used to treat a range of severe, chronic medical conditions resulting from hyperactivity of peripheral cholinergic neurons (see Chaps. 3–6 of this book). Indeed, such is the safety of BoNT, when used as a therapeutic agent, that its use has extended into cosmetic applications for the reduction of facial lines

K. A. Foster (✉)
Syntaxin Ltd., Units 4-10, The Quadrant, Barton Lane,
Abingdon Oxfordshire OX14 3YS, UK
e-mail: keith.foster@ipsen.com

K. A. Foster (ed.), *Clinical Applications of Botulinum Neurotoxin,*
Current Topics in Neurotoxicity 5, DOI 10.1007/978-1-4939-0261-3_1,
© Springer Science+Business Media New York 2014

caused by habitual facial muscle contractions, and it has Food and Drug Administration (FDA) approval for treatment of glabellar lines [2]. It is interesting to note that this dichotomy was apparent in the first scientific descriptions of botulism, long before the molecular identity of the causative agent was known. Following a number of outbreaks of "sausage poisoning" in Württemberg in south-west Germany, the district medical officer, Justinus Andreas Christian Kerner, published a series of papers between 1817 and 1822 that provide the first accurate and complete description of the symptoms of food-borne botulism [3]. Kerner extracted the active substance, which he termed "sausage poison" and believed to be a type of fatty acid, from the contaminated food and studied its effects, both in animals and on himself. Amazingly, in 1822, based upon these studies, he predicted the toxin's potential clinical utility, not only in conditions of muscular hypercontraction but also in autonomic conditions of glandular hypersecretion, such as hyperhidrosis and hypersalivation.

1.2   Therapeutic Use of the Neurotoxin

Following the original speculations of Kerner, it was not until 1980 that the first clinical application of BoNT was reported [4], [5]. Alan Scott was an ophthalmologist in San Francisco who in the late 1960s was seeking a chemical method to denervate striated muscle for the treatment of strabismus, a condition in which the eyes are not properly aligned with each other. Treatment of strabismus at that time was by surgical weakening of the extraocular muscles. The procedure was, however, unsatisfactory due to both high reoperation rates and its invasive nature. It had been shown in 1949 that BoNT blocks neuromuscular transmission through decreased release of acetylcholine [6], and the subsequent work of Drachman, a neuroscientist at Johns Hopkins University, demonstrated the effective denervation of skeletal muscles by local injection of BoNT [7]. Drachman introduced Scott to Ed Schantz, who was producing purified BoNT type A (BoNT/A) at the Department of Microbiology and Toxicology, University of Wisconsin and making it available for experimental use. Scott, working at the Smith-Kettlewell Institute, used the BoNT/A provided by Schantz in a series of experiments in monkeys that demonstrated its utility in weakening the extraocular muscles and its therapeutic potential for the treatment of strabismus [8]. Scott formed a small company, Oculinum Inc., in 1978 to develop BoNT/A for the treatment of strabismus, giving the drug the name Oculinum. Also in 1978, Scott received FDA approval to inject BoNT/A into human volunteers, and, in 1981, reported that botulinum toxin "appears to be a safe and useful therapy for strabismus" [9]. In 1988, the rights to distribute Oculinum were acquired by Allergan Inc., together with the responsibility to conduct clinical trials of its effectiveness for other indications, including cervical dystonia. In 1989, Oculinum Inc. received FDA approval to market Oculinum in the USA as an orphan drug to treat strabismus and blepharospasm associated with dystonia in patients 12 years of age and older. Shortly after the FDA approved these indications, Allergan Inc., acquired Oculinum

Inc. Allergan applied for and received FDA approval to change the product's name to Botox®.

Following Scott's seminal paper of 1981, clinicians and researchers around the world began to explore the therapeutic potential of this exciting molecule. Research showed the therapeutic benefits of BoNT/A extended far beyond ophthalmology, providing patients with temporary relief from facial spasms, neck and shoulder spasms, even vocal cord spasms, indeed all forms of focal dystonia proved suitable for treatment with local injections of BoNT/A (see Chap. 3 of this book). Ophthalmologists at the Moorfields Eye Hospital, London, were among the first clinicians in the UK to investigate the use of BoNT/A for correcting strabismus. They approached the Centre for Applied Microbiology and Research (CAMR) at Porton Down, a centre of excellence for research on botulinum toxins, to provide BoNT/A for their studies. CAMR developed a stable freeze-dried BoNT/A formulation for the Moorfields Eye Hospital which rapidly began to be used throughout the UK and in many European countries [10]. In 1990, the formulation was approved by the Medicines Control Agency in the UK for the treatment of blepharospasm and hemifacial spasm and marketed by CAMR's commercial partners Porton Products (now Ipsen Ltd.) as Dysport®. The name is short for Dystonia Porton Down.

The first report of a non-neuromuscular use of BoNT/A was by Bushara and Park [11] who, while treating patients with hemifacial spasm, discovered that BoNT/A injections inhibited sweating. BoNT/A was subsequently approved for the treatment of primary axillary hyperhidrosis and is now recognised to be an effective therapeutic treatment for a number of conditions involving hyperreactivity of the autonomic system (see Chap. 4 of this book). A major development in the commercial application of BoNT/A was the observation first reported in 1989 that it had an effect on wrinkles [12]. In 1992, the husband and wife team of JD and JA Carruthers (ophthalmologist and dermatologist, respectively) published a study on BoNT/A for the treatment of glabellar frown lines [13]. In 2002, Allergan obtained approval of Botox® Cosmetic, which resulted in a significant expansion of the commercial market, and in public awareness. By the end of 2006, Botox® sales significantly exceeded US$ 1 billion, with cosmetic uses accounting for about half of the sales. New applications for BoNT also continued to be developed and approved. Botox® received FDA approval for the treatment of chronic migraines on October 15, 2010 (see Chap. 6 of this book). Most recently, Botox® injections into the bladder were approved by the FDA: in 2011, for urinary incontinence due to detrusor overactivity associated with a neurologic condition in adults and, in January 2013, for the treatment of overactive bladder (OAB) with symptoms of urge urinary incontinence, urgency and frequency in adults (see Chap. 5 of this book).

The growth in the therapeutic application of BoNT combined with its cosmetic use has led to it becoming a major commercial product with huge growth potential. It has been estimated by analysts that by 2018 the total worldwide market for BoNT could be US$ 4.3 billion (Report by Global Industry Analysts, Inc.). This is all the more remarkable when it is considered that the dominant market products continue to be Botox®, which was the original commercial product launched in 1989, and Dysport®. It is also interesting to note that all of the major therapeutic products in the market are, with one exception, BoNT/A1 products. The one exception is a BoNT/B product registered with the US FDA in 2000 for the treatment of cervical dystonia by Elan Pharmaceuticals and subsequently sold to Solstice Neurosciences, LLC a wholly owned subsidiary of US WorldMeds, LLC who market the product as Myobloc® in the USA and as NeuroBloc® in the rest of the world. Myobloc®/NeuroBloc® is reported to have a higher incidence of side effects, particularly dysphagia and dry mouth, than type A products [14]. It has not gained a significant share of the BoNT marketplace. The majority of the BoNT products currently available, including both Botox® and Dysport®, are purified neurotoxin complex including neurotoxin-associated proteins (NAPs) as well as the neurotoxin protein itself (see Chap. 4 of the companion volume to this book, KA Foster (ed.) *Molecular Aspects of Botulinum Neurotoxin*, Springer, New York for an explanation of the neurotoxin complex and the role of the NAPs). In 2005, Merz Pharmaceuticals received approval in Germany to market Xeomin® which is a purified BoNT/A product without the NAPs. Approval for Xeomin® was given in Europe in 2007 and in USA in 2011. It is currently the only purified, noncomplex BoNT product in the market.

The concentration of BoNT products to the A serotype, and in particular the A1 subtype, is surprising given the diversity of BoNTs that have been identified. Not only are there the seven serotypes, A–G, but there are also multiple sub-types within each serotype now being identified, such that there are currently well over 30 subtypes of BoNT reported (see Chap. 10 of the companion volume to this book, KA Foster (ed.) *Molecular Aspects of Botulinum Neurotoxin*, Springer, New York). This diversity is growing as more clostridial strains are isolated and characterised. It is also now becoming apparent that the subtypes can differ significantly in their biochemical properties [15]. In one case, a subtype of BoNT/F, F5, has been reported to have a completely different substrate specificity to other BoNT/F subtypes [16]. This diversity represents a huge opportunity to identify new therapeutic neurotoxins with differentiated properties relative to the current products. The opportunity to expand the therapeutic landscape for BoNTs will be expanded even further by the recently established ability to create the neurotoxins by recombinant protein expression and to modify the neurotoxin protein to create engineered neurotoxins with enhanced and expanded therapeutic utility (see Chap. 7 of this book). Given the natural diversity of BoNTs and the ability to recombinantly express and engineer them, it is likely that the future growth in the clinical use of this remarkable family of proteins will exceed the current market expectations based upon the existing BoNT/A products.

1.3 Conclusion

In the 30 plus years since its first reported clinical use, BoNT has become a major therapeutic product. That remarkable development in the use of what is the most lethal toxin known is a reflection of the unique biology of this fascinating protein and the extensive research that has gone into understanding both the structure and function of this toxin over many years. The understanding that has resulted from those extensive studies combined with the application of the capabilities of the modern biotechnology industry promises that the exploitation of the BoNTs to the benefit of patients will continue to grow for many more years.

References

1. Arnon SS, Schechter R, Inglesby TV, Henderson DA, Bartlett JG, Ascher MS, Eitzen E, Fine AD, Hauer J, Layton M, Lillibridge S, Osterholm MT, O'Toole T, Parker G, Perl TM, Russell PK, Swerdlow DL, Tonat K (2001) Botulinum toxin as a biological weapon: medical and public health management. JAMA 285:1059–1070
2. Cheng CM (2007) Cosmetic use of botulinum toxin type A in the elderly. Clin Interv Aging 2:81–83
3. Erbguth FJ (2009) The pretherapeutic history of botulinum toxin. In: Truong D, Dressler D, Hallett M (eds) Manual of Botulinum toxin therapy. Cambridge University Press, Cambridge
4. Scott AB (1980) Botulinum toxin injection into extraocular muscles as an alternative to strabismus surgery. J Pediatr Ophthalmol Strabismus 17:21–25
5. Scott AB (1980) Botulinum toxin injection into extraocular muscles as an alternative to strabismus surgery. Ophthalmology 87:1044–1049
6. Burgen ASV, Dickens F, Zatman LJ (1949) The action of botulinum toxin on the neuro-muscular junction. J Physiol 109:10–24
7. Drachman DB (1971) Botulinum toxin as a tool for research on the nervous system. In Simpson LL (ed) Neuropoisons: their pathophysiological actions. Plenum Press, New York
8. Scott AB, Rosenbaum AL, Collins CC (1973) Pharmacological weakening of extraocular muscles. Invest Ophthalmol 12:924–927
9. Scott AB (1981) Botulinum toxin injection of eye muscles to correct strabismus. Trans Am Ophthalmol Soc 79:734–770
10. Hambleton P, Pickett AM, Shone CC (1981) Botulinum toxin: from menace to medicine. In: Ward AB, Barnes MP (eds) Clinical uses of botulinum toxins. Cambridge University Press, Cambridge
11. Bushara KO, Park DM (1994) Botulinum toxin and sweating. J Neurol Neurosurg Psychiatry 57:1437–1438
12. Clark RP, Berris CE (1989) Botulinum toxin: a treatment for facial asymmetry caused by facial nerve paralysis. Plast Reconstr Surg 84:353–355
13. Carruthers JD, Carruthers JA (1992) Treatment of glabellar frown lines with C. botulinum A exotoxin. J Dermatol Surg Oncol 18:17–21
14. Chapman MA, Barron R, Tanis DC, Gill CE, Charles PD (2007) Comparison of botulinum neurotoxin preparations for the treatment of cervical dystonia. Clin Ther 29:1325–1337
15. Henkel JS, Jacobson M, Tepp W, Pier C, Johnson EA, Barbieri JT (2009) Catalytic properties of botulinum neurotoxin subtypes A3 and A4. Biochemistry 48:2522–2528
16. Kal SR, Baudys J, Webb RP, Wright P, Smith TJ, Smith LA, Fernandez R, Raphael BH, Maslanka SE, Pirkle JL, Barr JR (2012) Discovery of a novel enzymatic cleavage site for botulinum neurotoxin F5. FEBS Lett 586:109–115

Chapter 2
Botulinum Toxin as a Clinical Product: Manufacture and Pharmacology

Andy Pickett

Abstract The use of botulinum neurotoxins as clinical products has required the development of suitable manufacturing processes meeting the highest standards of safety and quality for pharmaceutical products. The exceptional potency of the molecule has imposed safety requirements not just for the products but also for the staff and working environments during manufacture. There are now a number of clinical products available, produced using different manufacturing processes and with different final formulations. These differences and their implications are reviewed in this chapter. Future developments in the manufacture of clinical neurotoxin products are also considered.

Keywords Botulinum neurotoxin · cGMP · Drug substance · Drug product · Formulation · Containment · Pharmacology

2.1 Introduction

Botulinum neurotoxin (BoNT) has become widely accepted as a major, first-line treatment for a number of debilitating conditions that previously received no adequate therapy [1]–[3]. From the initial pioneering work of Alan Scott in ophthalmology [4], the use was expanded and extended into neurological conditions where muscles were in spasm [5], [6]. In the 20 years since the first licensed products became available—Oculinum® in the USA and Dysport® in the UK—BoNT has found established uses not just in neurology but in rehabilitation, urology, pain, headache, hyperhidrosis and aesthetic treatments [7]–[14], together with a wide range of new, experimental applications (see, for example, [15], [16] and Chaps. 3–6 of this book). In the majority of these indications, the products have provided long-term, stable results for patients [17]–[19].

A. Pickett (✉)
Toxin Science Limited, Wrexham LL12 8DU, UK
e-mail: ampickett@hotmail.com

Botulinum Research Center, UMASS Dartmouth, North Dartmouth, MA 02747, USA

K. A. Foster (ed.), *Clinical Applications of Botulinum Neurotoxin,*
Current Topics in Neurotoxicity 5, DOI 10.1007/978-1-4939-0261-3_2,
© Springer Science+Business Media New York 2014

Key to this predictable, long-term use, in many cases for the full duration of product availability over the decades, the provision of consistent, high-quality, reproducibly potent products has been essential [20]–[22]. Without this quality of product, clinicians could not have confidence that patients would have repeat, long-term therapy. Yet surprisingly, little has been published about how the BoNT products are manufactured, and data to show their consistency are scarce.

There are currently several manufacturers of type A BoNT for clinical use. Allergan, based in Irvine, CA, and Westport, Ireland, are the main manufacturers of the leading Botox®/Botox® Cosmetic/Vistabel®/Vistabex®/Vista® range. Ipsen, based in France and Wrexham, UK, are the second largest with the brands Dysport® and Azzalure® (an early brand name Reloxin® intended for aesthetic use in the USA [23] was finally not allowed by the US Food and Drug Administration, FDA). Merz, from Germany, have the Xeomin® and Bocouture® brands. Three other manufacturers are based in Asia. Medy-Tox, in Korea, have the Meditoxin®/Neuronox®/Siax® brand of type A BoNT which started in Korea and is now becoming available in other countries, for example, in Latin America. Hugel Inc., also based in Korea, have a further type A BoNT called Botulax (also Zentox or Regenox in other countries). Lanzhou Institute for Biological Products (LIBP), in Lanzhou, China, has produced a licensed type A BoNT available since 1997 called BTXA™ which is distributed by Hugh Source, in Hong Kong, and by other local/regional companies under different names such as Lantox in Russia; Lanzox in Indonesia; Prosigne in Brazil; Liftox in Ecuador; and Redux in Peru—in total, available in over 30 countries.

One serotype B BoNT has been available for a decade, called Myobloc®/Neurobloc®. The product was originally developed by Athena Neurosciences in the USA, subsequently acquired by Elan, then Solstice NeuroSciences LLC, now acquired in August 2010 by US WorldMeds based in Kentucky, USA. In Europe, the product is marketed by Eisai Europe Limited. Initially, the product was targeted to patients who had developed an antibody response to a type A product and so could be treated with type B BoNT since the two serotypes are immunologically distinct [24]. Follow-up treatment with immunoresistant patients has limited success in the long term [25]. Clinical data exist for licensed use of the product in cervical dystonia [26]–[28], but other clinical uses have been examined [25], [29]. However, large doses of the product are required, perhaps 50 times that for a type A product due to a lower potency per unit of toxin protein. Patients initially responded to this alternative but soon developed further antibodies to the type B molecules and became refractive to further therapy. The treatments were short-lived. Attempts have been made to revive the product as, for example, an aesthetic treatment [30]–[32] but the product is never used in routine aesthetic practice, despite these data. Of particular relevance to use of this serotype, recent data on receptor binding have shown that human synaptotagmin II, as used by type B BoNT, does not have a high affinity for the molecule [33]. This phenomenon is specific to humans and chimpanzees and potentially explains why high doses of type B BoNT are required to achieve therapeutic effects.

The present chapter describes how BoNT products are made for clinical use and the data available on those products, focusing on the three main BoNT-A families since these are the most prominent products in the marketplace today. The processes and their constraints are considered and how these link together is discussed. The pharmacology of the products is also discussed, especially with reference to actual clinical uses.

The mechanism of action of BoNT is described in detail elsewhere in this book (see Chaps. 2 and 6–8 of the companion volume to this book, KA Foster (ed) *Molecular Aspects of Botulinum Neurotoxin*, Springer, New York), and so is not included here. Also, clinical data are not considered in this section, unless applicable to the discussion.

2.2   The Two Stages of Manufacture

The manufacture of clinical BoNT is carried out in two, clearly distinct stages. All stages and all testing are performed in accordance with current good manufacturing practice (cGMP), which is a similar but not identical set of standards and requirements laid down by the regulatory authorities in each country and/or each region of the world. These standards will not be reviewed here, they are too extensive; the reader is referred to the freely available information of the guides and regulations on the web sites of the FDA and the European Medicines Evaluation Agency (EMEA) as typical examples of the stringency of manufacture and testing that are required.

2.2.1   *Bulk Active Toxin: Drug Substance*

The first stage is the production of the active BoNT in what are often described as 'bulk quantities'. In reality, exceptionally large amounts of highly potent active BoNT can be made in small, almost laboratory-scale equipment, unlike the type of quantities required for other active biomolecules. This involves cultivation of the production strain, initial separation of the crude BoNT, purification of the BoNT as either a complex or purified neurotoxin and final storage. Quality control (QC) testing of this material against a fixed and regulatory-approved specification is then carried out; in-process QC testing is also a stringent requirement. The resultant concentrated BoNT is called bulk toxin, bulk active substance, or, in the current modern terminology of the regulatory agencies, the active pharmaceutical ingredient (API) or the drug substance (DS). This multiplicity of terms is not so helpful; DS will be used here.

2.2.2   *Finished Product Vials: Drug Product*

After testing is satisfactorily completed and release of the DS for onward processing is approved by the quality functions in the company, very small volumes are used for dilution, formulation with excipients, dispensing into vials and then preservation by either freeze drying or vacuum drying or just as a liquid, dependent on the product family. The dilution factor may be 10,000 times or greater [34]. The output is called either finished product or, in current terms, drug product (DP).

2.2.3 Quality Testing and Release to Market

In certain cases, the regulatory authority of a country where a product is approved may require separate testing and release of the DS to be carried out, either by their own laboratories or by a designated and approved contract testing house, before any DP is manufactured. This can even be on an individual batch basis. In turn, each batch of DP may be individually released by the licensing authority of a country before that batch can be commercialised within the country. Within Europe, one control laboratory, designated an Official Medicines Control Laboratory (OMCL), will carry out formal testing and release on behalf of all European countries in accordance with the Official Control Authority Batch Release (OCABR) guidelines [35]. A fee is levied, currently about 2,300 € (US $ 3,200), against the manufacturer for the testing and release of each batch carried out together with the subsequent issue of a release certificate; but for the fee, time constraints are placed on the OMCL for release purposes. The OCABR for Human Biologicals is, in fact, a specific network within the OMCL network [36]. The FDA carry out similar testing for the products destined for the USA, in compliance with the Code of Federal Regulations (which define the cGMP requirements) 21CFR610 2(b), but no fee is levied. For other specific countries where products may be approved for sale, certification of testing and release from the home country of the manufacturer is required before the product batch is considered for commercialisation in the second country. For certain countries, repeat testing of any batch of DP submitted for commercial sale will be carried out by the respective authorities in that country, as part of the marketing authorisation when granted.

In most cases, the release authorities in a country carry out testing in accordance with the licensed testing procedures. This may go to extremes where, for example, even specific strains of mice are imported for the potency release testing of BoNT products. However, there are specific examples where a release authority may perform their own tests or even experimental tests as part of their release procedure. Those authorities maintain a rigid right to independent testing, to the extent that even the format and results from such tests are not revealed to the manufacturer. This has often been considered as unreasonable by manufacturers (and even other release authorities) to the extent that debate has often occurred on the reasonableness of such an approach. Changes have not, however, been brought about.

2.3 The Production Organism

2.3.1 Key Aspects

Clostridium botulinum is an obligate, spore-forming anaerobe, a gram-positive bacillus that is ubiquitous in nature. The nature of the bacterium leads to three distinct aspects that must be dealt with for the production of clinical material to occur successfully.

Firstly, the anaerobic requirements mean that oxygen must be excluded from the first stages of the production system, when the bacteria are grown in a fermenter or a glass vessel. A fermenter is the preferred usual choice in these cases, since this may be steam sterilised under pressure and fully validated in accordance with modern regulatory requirements and cGMP. The culture media employed generally include a reducing agent to aid this anaerobic requirement, and additional nitrogen may be used as an overlay. During growth, the bacteria produce several volatile acids that help to maintain an anaerobic environment and give the characteristic odour of a fermenting culture!

Secondly, the production of toxin progresses from the first stages of growth [37], only recently reported for a strain used in a clinical product process [38]. Suitable precautions to ensure that the culture fluids are appropriately contained and handled correctly, to prevent contamination of the environment or the operators, are essential. Gases evolving from the growth stages will build up pressure in the growth vessel and therefore must be filtered before release to prevent bacterial aerosols escaping.

Finally, sporulation of the bacteria can occur at low levels during the growth stages, but particularly in abundance as the bacterial life cycle ends and the bacteria die. A high spore count in the final stages of the fermentation is inevitable [39]–[41]. Non-sporulating mutants of type A or B strains have not been reported but new mutants with reduced sporulation have recently been generated; these also have reduced BoNT production [42], indicating the two processes may be linked genetically. The absence of spores has been reported for one production strain, but this may simply be due to the fermentation conditions, including the medium, employed [43]. Stages therefore need to be added to the manufacturing process to ensure that no spores are present in the final active BoNT DS used for clinical product, usually by filtration. The spores are highly resistant to heat and radiation and the use of these techniques is not possible.

The nutritional growth requirements of *C. botulinum* are not known in detail. Perhaps one of the most comprehensive reviews of growth and toxin production was described in a series of papers in the late 1950s and early 1960s by Bonventre and Kemp [37], [44]–[46]. Complex growth media are therefore used. As these can originate from plant and/or animal sources, the provenance must be established before use. Only certified sources of animal materials that are free from adventitious agents must be used. Even plant material derivatives can be processed by the suppliers using enzymes of animal origin and so careful investigation of the sources and the processes used for the materials is essential before purchase. One BoNT manufacturing process described for clinical material, essentially the original Botox® process, has used no materials of animal origin [47]. Inevitably, BoNT product manufacturers closely guard the finer details of their processes, especially production conditions used, as their intellectual property and know-how. There is certainly evidence to suggest that the process described by Schantz and Johnson [47] has now changed [48].

2.3.2   Strains Used

The type A BoNT production strains are all from the original collections of Professor Ivan Clifford Hall, a distinguished and enigmatic anaerobe specialist and epidemiologist of botulism food poisoning, amongst several of his specialities. Hall worked in the first half of the 1900s and collected many strains of *C. botulinum* from all over the USA. For an unknown reason(s), many BoNT type A production strains became known as 'the Hall strain', which is inaccurate [49]. Hall was known to have several type A strains in his collection, even in the earliest days of his work [50]. Despite this, one manufacturer has recently stated that type A products are from the same strain [51], which is again inaccurate [20]. In fact, recent genetic analysis of type A strains, including clinical product strains, has shown them to be clearly distinct from each other [52].

Detailed genetic analysis of the neurotoxin gene within the Allergan production strain has been published [53], [54] and complete genome sequencing of another strain, designated ATCC 3502 and as used by Merz [51], has been available for several years [55]. The exact strains used by Ipsen and MedyTox have not been published, other than being described as 'a Hall strain' [34], [56]. The neurotoxin sequence of the strain used for the production of BTXA by Lanzhou Institute for Biological Products has also been published [57].

The BoNT type B production strain used is designated the Bean strain. The production method was originally described in detail by Siegel and Metzger [58], which was updated for the clinical product by Setler 20 years later [59].

To date, only BoNT serotypes A and B have been made available for clinical uses. Other serotypes have been tested in man [60]–[64], but none have yet become commercial products and have progressed no further than proof-of-principle tests in man.

2.4   The Critical Mixture of Safety and Good Manufacturing Practice

There are two main challenges that override the manufacturing processes for clinical BoNT production.

The keyword to drive the processes is 'containment'. Due to the high potency of BoNT and the sporulation abilities of the production organism, a detailed but risk-based approach to safe operation must be taken to protect both the operators and the environment. The production environments will normally be high-grade rooms with special air-handling systems and these must not become contaminated or lengthy decontamination procedures will need to be followed that ensure (and are validated to ensure) the elimination of any spores, bacteria and BoNT—all of which must be considered separately. Only limited data are available on, for example, the persistence of BoNT after decontamination of surfaces [65]. Information on typical methods of decontamination by formaldehyde or other fumigants is also available and useful [66], [67].

Coupled with the safety aspects, manufacture must proceed in accordance with the cGMP requirements of the national or regional licensing authorities from whom marketing authorisations are sought for the products. These are laid down in detail and few, if any, exceptions are granted by the authorities because of the nature of either the product or the organism. This is especially the case since the current manufacturers have been compliant with these standards for many years, firmly setting the quality goals that any new manufacturer must meet for market entry.

The safety and containment requirements for facilities handling human pathogens such as *C. botulinum* are defined in three documents which are widely available as the basis for the standards. These are from the World Health Organisation [68], the United States Health and Human Services (HHS; Center for Disease Control and Prevention, CDC) [69] and the United Kingdom Health and Safety Executive Advisory Committee on Dangerous Pathogens [70]. Each of these contains descriptions of the standards to be applied. Within each country, different organisations will also be involved in the legislative and standards process—for example, CDC is the branch of the HHS responsible for both *C. botulinum* and, separately, for work with BoNT, including handling, storage and shipping. Amounts of BoNT up to 0.5 mg are not regulated under current US legislation (see Code of Federal Regulations 42CFR73.3 ((d)(3))).

The *C. botulinum* organism is currently considered in somewhat different ways by different countries. In the USA, BoNT (the toxin and not the producing organism) is one of six Category A Bioterrorism Agents defined by CDC. The definition of such an agent is as follows [71]:

High-priority agents include organisms that pose a risk to national security because they

- *Can be easily disseminated or transmitted from person to person;*
- *Result in high mortality rates and have the potential for major public health impact;*
- *Might cause public panic and social disruption and*
- *Require special action for public health preparedness.*

Both BoNT and the bacterium are however included on the select agent list in the USA [72] and in 42CFR 73.3. A detailed publication on what is required for the security of such agents has been prepared by both CDC and the US Department of Agriculture [73] and describes all aspects of a facility security policy and procedure, from personnel access to chain of custody requirements. These are additional to the other biosecurity measures that have to be taken for safe handling and processing, as described earlier.

In the UK, *C. botulinum* (*not* BoNT) is defined as a hazard group 2 biological agent whereas, in contrast, *Bacillus anthracis* is in hazard group 3 [74]. The list also indicates that a vaccine is available, but vaccination of individuals involved with the organism also has potential risks associated with the vaccine itself. In fact, the only commonly available vaccine has been withdrawn from use mainly due to efficacy issues [75]. New generation BoNT vaccines are being actively developed [76], [77]. Detailed risk assessments on vaccination are therefore called for before its adoption as a standard procedure in the production environment. Availability of a suitable vaccine remains a key issue if any company risk assessments indicate that vaccination is a vital part of their production strategies.

To meet these dual requirements, manufacturers generally can adopt two approaches. Firstly, mainly due to the significant quantities of BoNT being handled and the high concentrations of the production organisms, which are quite unlike those found in a normal pathology or microbiology environment such as a diagnostic laboratory, manufacturers can uprate the pathogen-handling categorisation for their facilities to the next level or higher. This enables a high degree of re-assurance that handling of the organism and toxin is under the strictest containment procedures. Secondly, an increase in the categorisation means that the work should be performed in class III microbiological safety cabinets, which are fully contained, negative pressure units with glove-port access only (Fig. 2.1). The exhaust air from these cabinets is then high-efficiency particulate arrestant (HEPA) filtered through monitored and tested specific grades of filter specifically for environmental protection. Occasionally, pictures of these production cabinets from manufacturers are shown at presentations in symposia or in their product literature, but, apart from these, nothing else is published about the technology involved.

2.5   The Production Processes

There are many references to procedures and processes for the manufacture of BoNT, covering several decades (see, for example, [47], [78]–[81]), but until the 1990s, these of course did not focus on (or even envisage) any clinical uses and hence clinical quality of product. Nevertheless, these original processes formed the basis for clinical product manufacture that subsequently occurred.

2.5.1   *Botox*® *Family*

The only detailed publication that describes manufacture of clinical products is that of Schantz and Johnson, in the early 1990s [47]. As the originators of Oculinum® (later to become Botox®), they published in detail the exact process used for the manufacture of DS at that time (Fig. 2.2), which, as described, was based on the original method of Duff and colleagues [82], [83]. In particular, the method described was that used for DS batch 79-11, made in November 1979. This batch was critical in the clinical trials for the product and, subsequently, for the launch of commercial product. The batch lasted for many years after manufacture and was distributed as Botox® by Allergan until 1997 [84] (approval by FDA of alternative DS) or later in other markets outside the USA. This was despite the fact that batch 79-11 was reported as losing potency over the nearly 20 years since manufacture [47]! The process was 'classical', involving the original first stage of BoNT precipitation with an acidification step, followed by several stages with salt solution, ethanol precipitation

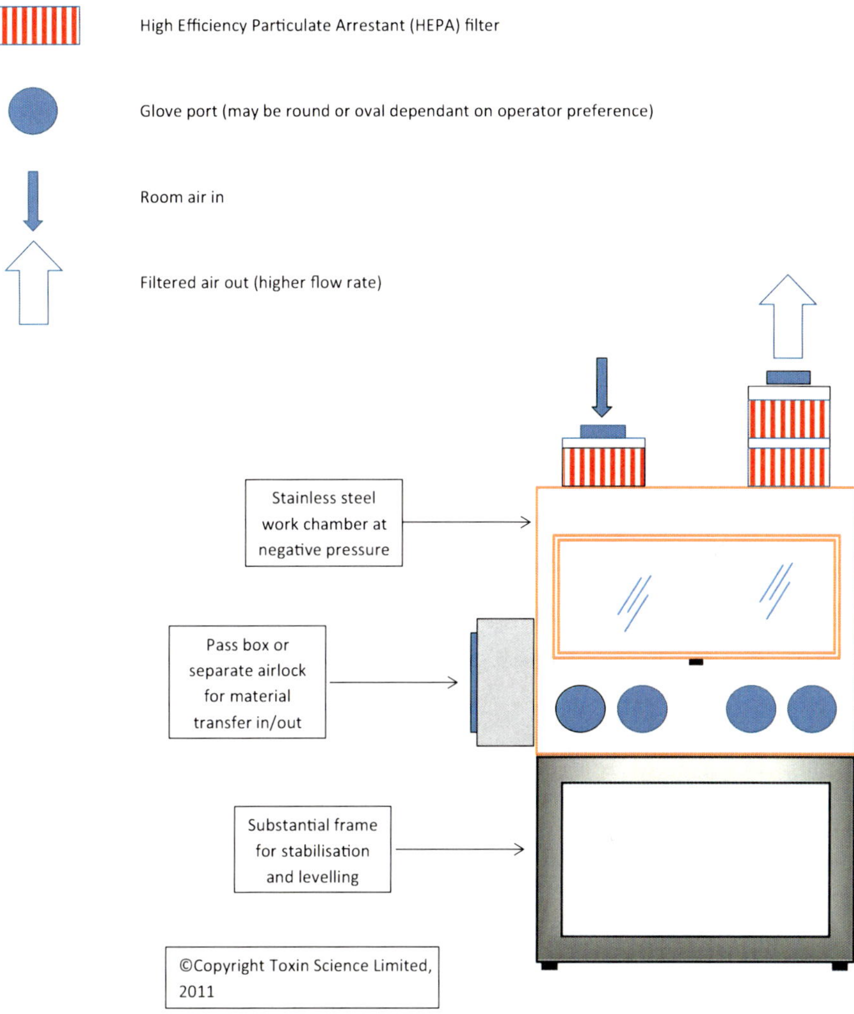

Fig. 2.1 Schematic drawing of class III microbiological safety cabinet used for manipulation of high-potency/volume BoNT-containing materials

and crystallisation in the presence of ammonium sulphate [47]. A clear specification for the DS was established (Table 2.1), as they described [47]. Interestingly, the specification included a measurement of the amount of residual nucleic acids present, as a proportion of the protein obtained (expressed as an $A_{260}:A_{280}$ ratio), which indicated that these were considered contaminants at that time. The DP was stabilised with human serum albumin (HSA) in a formulation that was intended to be physiological in properties, with sodium chloride as the second component [47].

Some years later, in 2008, a further paper—this time published by Allergan—claimed that the method was still used to manufacture the Botox® DS [85], [86]. However, in between times, an inspection report of an Allergan-contracted

　　　　　　　　　　　　　　　　　　　　　　　　A. Pickett

Production Scheme for Oculinum / Botox
Batch 79-11 (200mg)

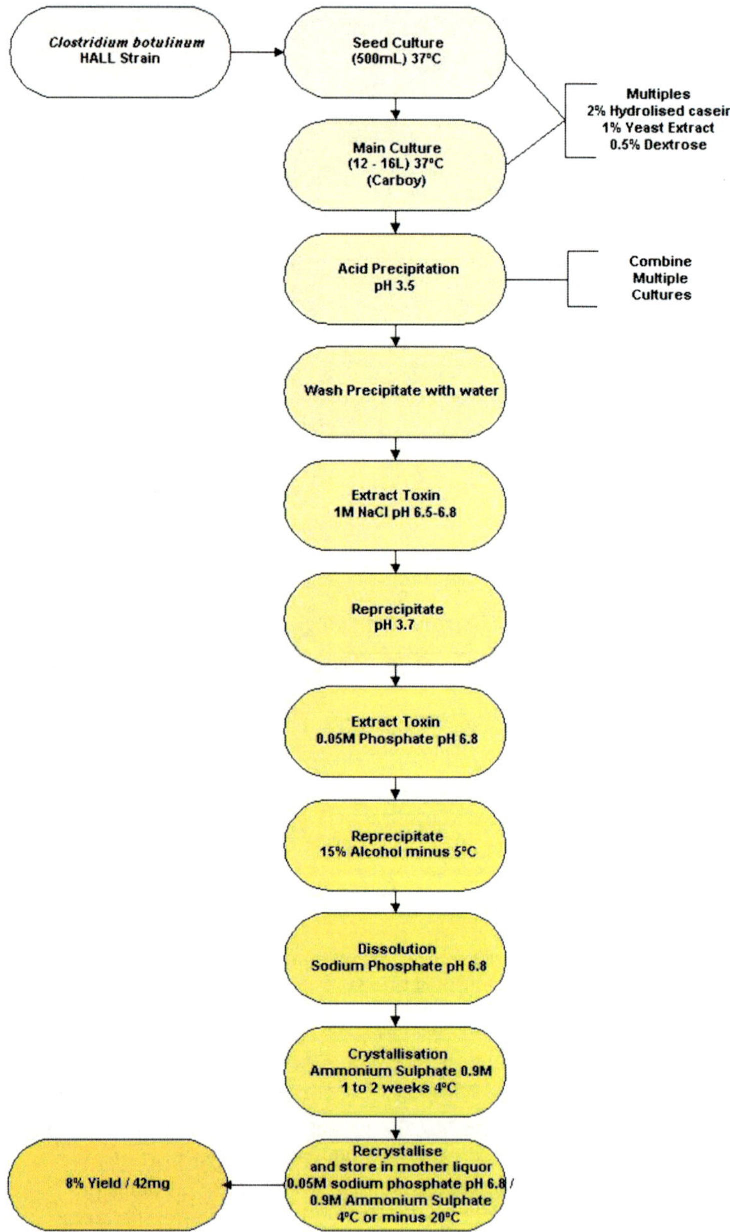

Fig. 2.2 Production scheme employed for manufacture of Oculinum®/Botox® batch 79–11, the original product first licensed in the USA in 1989. (Scheme based on details in reference [44])

Table 2.1 Drug substance (bulk active BoNT) specification for original Oculinum®/Botox® produced by Edward Schantz for initial clinical trials and subsequent product licensure. (Based on reference [47])

Product parameter	Specification
Strain	Hall
Absorbance (in 0.05 M sodium phosphate buffer pH 6.8)	Maximum at 278 nm
A260/A278 nm ratio	0.6 or less
Specific toxicity (mouse potency)	$3 \times 10^7 \pm 20\%$ mouse LD_{50} per mg
Extinction coefficient (absorbance) 1 mg per ml toxin in 1 cm light path	1.65

establishment [87] indicated that the process had changed in various aspects from the original Schantz and Johnson [47] method and this was raised in the follow-up correspondence [48]. No direct response was received to these questions and the subject has not been addressed since. Although Allergan produced DS at this contractor for a time, the current production of DS is believed to take place in Irvine, Orange County, CA with finished DP originating from the Allergan Westport facility, in Ireland—as noted on the labelling in most countries for the product family.

2.5.2 *Dysport® Family*

Two papers have summarised how the Dysport® family of DP are manufactured [34], [88]. The process is described as 'precipitation, column chromatography and dialysis'. The use of column chromatography in the Dysport® DS process was in fact criticised by Schantz and Johnson in their main paper [47] as something that should be avoided, but the Dysport® experience has shown and demonstrated the validity of using such methods of production.

Many attempts have been made by authors in the past to cite a specific reference [80] as describing how the Dysport® family of products is manufactured. However, these attempts have been repeatedly criticised by the manufacturer (see, for example, [89]). The publication cited was nearly 10 years before Dysport® was licensed as a therapeutic product at the end of 1990. The errors in such claims have been detailed [89]. A second publication [81] has recently been cited as again describing the production process [51], but there are very few details therein, which only mention an eight-stage process with 46 in-process and quality control tests using a 30-litre stainless steel fermenter. There is no literature or other information to support if these data are accurate for the Dysport® process today.

These incorrect citations have, in turn, led to errors in published product-specific data, especially product characteristics, which has not been helpful to clinicians [89]. These errors have also included data on DP formulations [21]. Consequently, Pickett and Perrow published in 2010 [90] a definitive description of the current BoNT formulations available in the market, with DS characteristics included (Table 2.2). Regrettably, reviews are still being published, at the time of writing the present chapter, that include erroneous product data [21], [91] and which do not even cite the source of the data they used.

Table 2.2 Characteristics of current major BoNT products. (©2012 Toxin Science Limited. Reproduced with permission)

Product	Production strain	Process	U/vial [Product specific]	Excipients (in vial)	
Dysport®	Hall	Fermentation Dialysis Chromatography	300/500 sU	HSA 125 ug	Lactose 2.5 mg
Azzalure®	Hall	Fermentation Dialysis Chromatography	125 sU	HSA 125 ug	Lactose 2.5 mg
Botox®	Allergan "hyper"	Fermentation Precipitation "Crystallisation"	100/200 B	HSA 500 ug	NaCl 0.9 mg
Vistabel® & Vistabex®	Allergan "hyper"	Fermentation Precipitation "Crystallisation"	50 V	HSA 500 ug	NaCl 0.9 mg
Xeomin®	Hall	[Unpublished]	100/200 X	HSA 1 mg	Sucrose 5 mg
Bocouture®	Hall	[Unpublished]	50 B	HSA 1 mg	Sucrose 5 mg

2.5.3 Xeomin® Family

No publication exists to describe the manufacture of the Xeomin® product family. Even a recent publication, claiming in the title to consider manufacturing of the product, does not [92]. As for the facilities and equipment used, limited data have been presented in congresses on the manufacturing process and this seems to follow other processes used previously for purified BoNT. In addition, data included in patent applications by the scientists who developed the product have contributed towards the process data available [93]. This source is, however, somewhat restrictive and not mainstream for clinicians seeking further information on this product family. Repeated requests have been made for such data to be published and hence made available to the clinical community, but this has not yet happened.

2.5.4 *Myobloc®/Neurobloc®*

There are publications available which describe how the type B BoNT product is manufactured [59], [94], [95]. The process applied is similar to that for BoNT type A except that an additional stage of BoNT activation is required for full activity to be obtained. [59], [95]–[97]

2.6 Consistency and Quality of Products

For clinicians to be able to judge the merits of the different products available, clear and unequivocal product data need to be made readily available [21], [91]. Aspects such as consistency of products manufactured over time, stability, biochemical characteristics and potency are key to assist in the comparisons. Such data are, however, very limited for reasons that are not entirely clear.

Because of the product licensing procedures and the subsequent DP quality release procedures by the licensing authorities, consistency of final products should be inherent, or the products would not be available for marketing. These are facts. However, data should be available to the clinical community in order to judge product consistency for themselves.

2.6.1 *The Initial Oculinum®/Botox® Issues*

When Oculinum® was first licensed by the US FDA, all the product originated from one batch of bulk toxin [47]. This batch was labelled as 79-11 being made in November 1979. The batch lasted for many years of commercial production and sales, but was found to have significant issues about potency [47], [98]. Approximately 90 % of each finished product vial was found to be inactive BoNT, material that had been inactivated by the final stages of the manufacturing process for the vials [83], [98]. For patients, this meant that they were being given some ten times more BoNT-related protein than needed to achieve the therapeutic effect. As a consequence, many patients developed neutralising antibodies to BoNT protein, especially when being treated with the higher doses used for conditions such as cervical dystonia or stroke-related paralyses. This was recognised in the early 1990s [99]–[101]. In addition, the treatment regimens at that time often used so-called booster injections, where patients would be retreated after a short time interval (usually weeks) if their initial response was not felt to be sufficient [102]. Up to 17 % of patients with cervical dystonia were reported with these antibodies [103] and hence became refractory to BoNT therapy. These data have been revised downwards more recently, based upon long-term experience with the current Botox® product [104], [105]. Interestingly, this issue did not seem to occur as much with patients outside the USA [83] as another batch of Botox® DS was apparently used to supply those markets.

These issues relating to the product were reported by the originators of the BoNT DS [47], [83], but not by Allergan. The originators also identified ways in which the inactivation in the DP manufacturing process could be eliminated and proposed new formulations that might be considered which retained almost the entire potency [98].

In the event, Allergan chose to replace the DS batch with a new designated BCB-2024 [84]. New preclinical testing was carried out [84], followed by clinical studies in cervical dystonia. Initial comments were that the new DS batch was of significantly higher potency and caused side effects in patients, but later reports recorded equivalence in efficacy and safety profiles with the original [106]–[108]. FDA finally approved the new DS in 1997 [84]. There was no change in the formulation composition of the DP, but no details have ever been made available as to whether the actual DP process was changed or not, in line with the recommendations from the originators of the product [83], [98].

These initial issues with Botox®, effectively a significant inconsistency of the product to yield consistent clinical results over time, have not occurred again. Allergan have only published limited analytical data on the later and current DS materials used [48], [85], [86], but have offered no reassurances to the clinical community that such issues will not occur again. Allergan does, however, appear to continue to focus on the science related to neutralising antibody formation [109]–[111] and has recently published a meta-analysis on the use of Botox® in various clinical indications, which demonstrates an overall low level of antibody response to the product when used in various clinical applications [105]. Today, the number of patients developing neutralising antibodies to the product is no greater than for any of the other type A BoNT products [112].

2.6.2 Consistency of the Dysport® Product Family: The History of Protein Load

In 2003, Ipsen made available the first data on the consistency of Dysport® over time [113]. These data were published due to an argument that had developed between the manufacturers as to how much toxin-related protein (both neurotoxin and complex-related proteins) was there in a vial of each product. At that time, the Merz products were not available and the debate was between Dysport® and Botox®.

Because of the number of patients who had developed antibodies to early BoNT therapies, the subject of 'protein load' had emerged. Protein load was defined as the amount of all toxin-related protein per vial of product—both neurotoxin and complex-related proteins (but excluding the HSA excipient) and was represented in nanograms. Giving a patient the least possible amount of toxin-related protein was considered one way of minimising the potential for that patient to develop antibodies and hence become 'resistant' to treatment [114].

One early reference had cited a predecessor product to Dysport® as containing 12.5 nanograms protein load per vial [115]. Conversely, the data available for Botox® had indicated that this only contained 5 nanograms per vial [84], [108], [116]. The argument was established.

On review, the Stell reference [115] was found not to describe Dysport® and this was documented several times by Ipsen scientists [34], [117], [118]. In evidence, they produced in total four publications [34], [113], [117], [118], the last of which in 2008 [34] demonstrated a 15-year consistency of product data and only 4.35 nanograms of toxin-related protein per vial. On a per treatment basis (taking into account the potency unit differences), this meant that patients treated with Dysport® only received 40–50 % of the toxin-related protein that patients treated with Botox® were given. The 2008 data also demonstrated DS consistency over the life of the product in detail and reported biochemical properties including binding and enzymatic activity. Overall, the product was shown to be of high consistency and quality, with a number of DS batches being used throughout the product life cycle.

The consistency of Dysport® DP has also been demonstrated in a non-clinical model of muscle force in the rat (the rat muscle force (RMF) model) [119]–[121]. The model, based on one reported earlier [122] and subsequently developed further by Ipsen in conjunction with the University of Florida, is clinically relevant since muscle force is temporarily weakened in a dose-dependent manner using doses that are similar to those used in humans (on a per kilo basis). The data provided were reproducible and useful as measures of consistency (Fig. 2.3). The model was used to show consistency of Dysport® DS made in three different facilities. Unlike other non-clinical animal models that have been used for BoNT products, such as the digital abduction scoring (DAS) method [84], [123], the RMF method used a quantitative, non-subjective assessment of the effect of BoNT, in comparison with a placebo.

One further source of consistency data exists for the Dysport® family in the publically available review reports from the FDA which have been made available subsequent to product licence approval in the USA. In the clinical sections of these reports, data are provided of a trial in humans with cervical dystonia which used DS sourced from two different manufacturing sites (trial reference Y-97-52120-096 [124]). These are highly relevant consistency data since they originate from man. The data show clearly that no differences in clinical response were found using material from either source. Essentially, this demonstrates the robustness of the DS manufacturing process, since transfer between manufacturing sites can be a severe test of the reproducibility of any biological manufacturing process. An earlier description of biological product manufacture used to be 'the process defines the product', a short way of saying how process-dependent biological products used to be.

As Botox® was licensed in the USA many years prior to Dysport®, no such publically available data exist from FDA records. One publication has described the licensing process for Botox® [125] and the somewhat abbreviated dossier that was originally filed and approved.

2.6.3   The Xeomin® Family

No data exist in the public domain on the consistency of the Xeomin® product family.

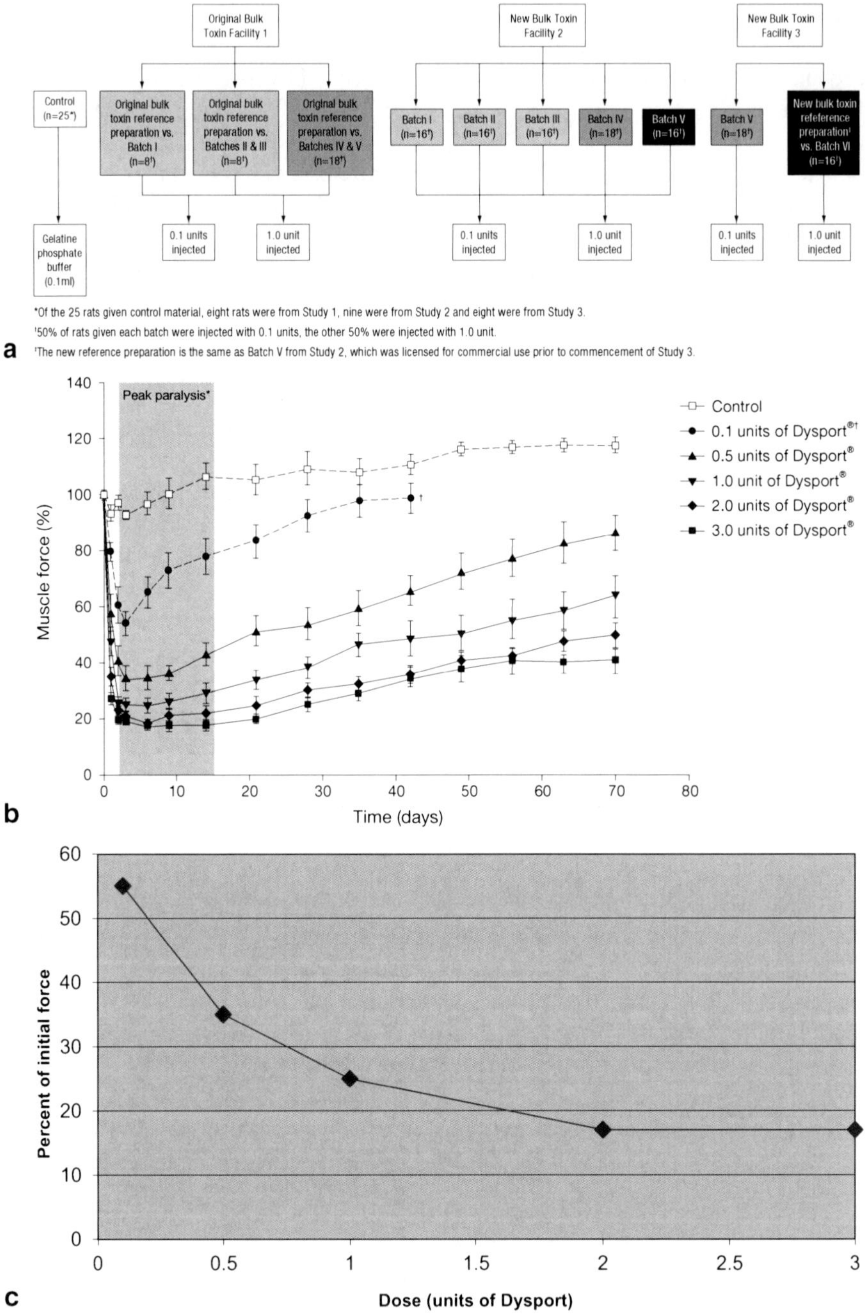

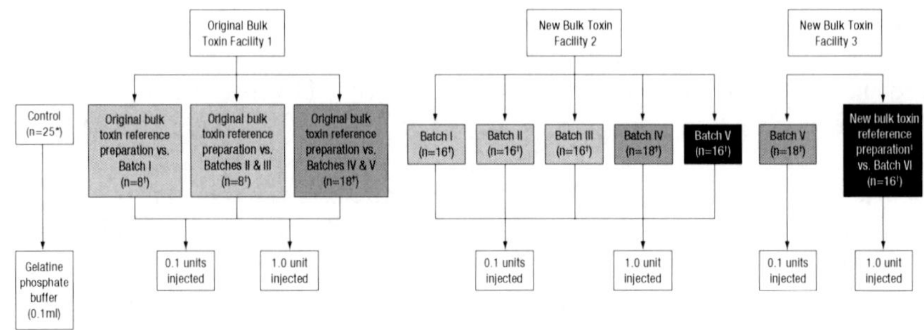

*Of the 25 rats given control material, eight rats were from Study 1, nine were from Study 2 and eight were from Study 3.

†50% of rats given each batch were injected with 0.1 units, the other 50% were injected with 1.0 unit.

a ‡The new reference preparation is the same as Batch V from Study 2, which was licensed for commercial use prior to commencement of Study 3.

b

c

Fig. 2.3 Use of the rat muscle force (RMF) model for the assessment, characterisation and comparison of clinical batches of Dysport® botulinum toxin type A. **a** Detailed schematic diagram of how the product batches from different manufacturing sources were tested using the method.

A useful publication, providing detailed non-clinical data and characteristics, was made available early on in the product's life (when the product was still known by the Merz compound number NT201) [126]. This did not, however, provide DS or DP consistency data.

Unusually, one clinical paper exists which discusses what might be an early *inconsistency* of Xeomin® in clinical use [127], which was further tested by the reporters in additional clinical studies of glabellar line treatment. No continuing problems were identified and the original issues identified were not repeated.

2.7 Stability of BoNT Products

The subject of stability of the BoNT products has evolved since the products were first licensed. Table 2.3 shows the licensed storage conditions that are granted now, for the three main product families.

2.7.1 Shelf-Life Storage

Originally, Botox® had to be stored below $-5\,^{\circ}\mathrm{C}$ whereas Dysport has always been maintained at refrigerator temperatures of $2\text{--}8\,^{\circ}\mathrm{C}$. Botox® now has a mixed set of storage conditions, either below $-5\,^{\circ}\mathrm{C}$ or at $2\text{--}8\,^{\circ}\mathrm{C}$. Xeomin storage is below $25\,^{\circ}\mathrm{C}$ and this has not changed since the product was first licensed. For biological products, the duration of shelf-life storage for Botox® and Dysport® has increased until the maximum of 3 years and 2 years, respectively. Of 67 biologics examined recently for length of shelf life, licensed over an 18-year period, none had a shelf life longer than 3 years [128].

Since first arriving on the market, Merz has made much of the elevated storage temperature of Xeomin® and has published storage stability data several times [129], [130]. In different publications (see, for example, [130]), storage for short periods at temperatures up to $60\,^{\circ}\mathrm{C}$ has been reported. The stability of the product is excellent under such conditions. Why? The Xeomin® family has the highest concentration of HSA excipient per vial (Table 2.2), eight times that found in Dysport® and twice that of Botox®. HSA is an exceptionally stable human protein. The HSA-manufacturing process has always included a step of heating at $60\,^{\circ}\mathrm{C}$ for 10 hours as one of the viral

---

b RMF generation results from a study to identify suitable doses for future studies in the model. The period of peak paralysis is represented by the band of shading. In this case, peak paralysis is defined according to specified acceptance criteria (days 2–15). The 0.1 unit dose was used to assess duration of effect. When muscle force generation returned to 100 %, no further measurements were made. **c** Dose–response relationship for Dysport® product batches tested in the RMF model. Peak paralysis response values highlighted in (b) were used. (Reprinted from Pickett A, O'Keeffe R, Judge A, Dodd S (2008) The in vivo rat muscle force model is a reliable and clinically relevant test of consistency among botulinum toxin preparations. Toxicon 52(3):455–464, Copyright (2008), with permission from Elsevier)

Table 2.3 Current storage conditions licensed for the different BoNT product families in major regions of the world. (Data taken directly from the official information provided by each regulatory authority web sites or official product web sites in each region)

Product[a]	EU	USA	Japan	Australia
Dysport® (300 and 500 Speywood units)	2 years at 2–8 °C	12 months at 2–8 °C; protect from light	N/A	2 years at 2–8 °C
Azzalure® (125 Speywood units)	2 years at 2–8 °C	N/A	N/A	N/A
Botox® (50, 100 and 200 units)	3 years at 2–8 °C or at/below −5 °C	3 years at 2–8 °C (100 units) or 2 years at 2–8 °C (200 units)	Below 5 °C (stability data for 50 and 100 unit sizes is provided for 2 years)	2 years at 2–8 °C (200 units) or 3 years (100 units)
Botox® Cosmetic (50 and 100 units)/Vistabel®/Vistabex® (50 units)/Botox Vista® (50 units)	3 years at 2–8 °C (Vistabel®/Vistabex®)	3 years at 2–8 °C (Botox® Cosmetic)	Below 5 °C (Botox Vista®)	N/A
Xeomin® (50 and 100 units)	4 years at 25 °C or below (100 units) 3 years at 25 °C or below (50 units)	Up to 3 years at room temperature 20–25 °C (68–77 °F), in a refrigerator at 2–8 °C (36–46 °F), or a freezer at − 20 to − 10°C (− 4 to 14 °F).	N/A	N/A
Bocouture®	3 years at less than or equal to 25 °C	N/A	N/A	N/A
Myobloc®/Neurobloc®	3 years at 2–8 °C; protect from light	3 years at 2–8 °C; protect from light	N/A	N/A

N/A product not available in that country
[a]Not every product unit size is available in every country

inactivation stages used to provide assurance of freedom from human adventitious agents [131]. The higher HSA concentration in Xeomin® may, therefore, confer high temperature stability for the product, probably acting as a protecting protein or similar. No explanations have been forwarded by Merz for this effect. The standard storage conditions of the Xeomin® family, at less than 25 °C (perhaps less than 30 °C in some countries, such as Brazil), cover many shipping conditions, but in higher temperature climates and summer conditions this will be exceeded, necessitating shipping under controlled temperature conditions, as for the other BoNT products. The higher storage temperature for Xeomin® does mean that refrigeration is not required by most pharmacies for the product.

Certain of the newer low-unit formulations, such as Azzalure® or Vistabel®, have the shelf life of their parent. As each product has been submitted for licensing, individual stability data also have to be submitted to the regulatory authorities for their review and approval. In cases where the shelf life is shorter than the parent product, this is likely due to the new product only having stability data to support the initial shelf life at the time of licensing. This does not mean the new product is 'less stable' or of 'lower quality' than the parent and no such conclusions should be drawn. In due course, these shelf lives will be extended.

2.7.2 Reconstitution Stability

The stability of the BoNT products after reconstitution, using the recommended diluent of sterile 0.9 % sodium chloride, has also been an area of commercial competition. The post-reconstitution time is variable depending on the requirements of the individual licensing authorities. Details of several licensed conditions are shown in Table 2.4.

As more data have been generated by the manufacturers, they have been able to gain extensions to post-reconstitution storage conditions until now when the maximum post-reconstitution storage time has been reached in the various markets.

Clinicians regularly ask the manufacturers to supply information on reconstitution stability. In many cases, especially in the aesthetic world, a long reconstitution time is sought, beyond that stated in the standard supplied product information. The manufacturers have only been able to supply limited information, basically restating what conditions have been approved by the licensing authorities occasionally supplemented by small additional studies that they have performed to provide data if, for example, excursions in storage condition occur. Consequently, clinicians have performed numerous studies of their own to look at, for example, extended reconstitution times or even alternative diluents. These studies have provided valuable data for the clinical community and should be taken into account in any discussions relating to the individual products. A summary of such studies is provided in Tables 2.5 and 2.6.

These studies have, importantly, looked at maintenance of clinical effects after long-term reconstitution, together with the microbial content (that is, sterility) of the reconstituted product—one of the main risks, since the products are essentially a protein-rich medium (with their content of HSA) for microbial growth after reconstitution. Typically, the studies by Hexsel and co-workers on Dysport® [132] and Yang and colleagues on Botox® [133] have been significant contributors to data on reconstituted products.

Table 2.4 Current reconstitution storage conditions licensed for the different BoNT product families available in different regions of the world

Product	EU	USA	Japan	Australia
Dysport®	Up to 8 h at 2–8 °C following reconstitution	Up to 4 h at 2–8 °C following reconstitution; protect from light	N/A	Up to 8 h at 2–8 °C following reconstitution
Azzalure®	Up to 4 h at 2–8 °C following reconstitution	N/A	N/A	N/A
Botox®	Up to 24 h at 2–8 °C following reconstitution	Up to 24 h at 2–8 °C following reconstitution	Use immediately	Up to 24 h at 2–8 °C following reconstitution
Botox® Cosmetic/ Vistabel®/ Vistabex®/Botox Vista®	Up to 4 h at 2–8 °C following reconstitution (Vistabel®/ Vistabex®)	Up to 24 h at 2–8 °C following reconstitution (Botox® Cosmetic)	Use immediately (Botox Vista®)	N/A
Xeomin®	Up to 24 h at 2–8 °C following reconstitution	Up to 24 h at 2–8 °C following reconstitution	N/A	N/A
Bocouture®	Up to 24 h at 2–8 °C following reconstitution	N/A	N/A	N/A
Myobloc®/ Neurobloc®	If diluted, use immediately	If diluted, up to 4 h at 2–8 °C; protect from light	N/A	N/A

Reconstitution conditions often state that from a microbiological point of view, the product should be used immediately after reconstitution

N/A Product not available in that country

2.8 New Product Formulations

The current BoNT product formulations have existed, for the main products, for more than 20 years. In that time, they have changed in no respect. They have shown themselves to be stable (see above), convenient for use and excellent for patient treatments. The number of reports of patients developing adverse responses immediately or shortly after injection is limited, taking into account the usage of the products across the world now [134]. Only recently, with the advent of aesthetic uses, have alternative versions become available, but these have only reduced the number of BoNT units per vial and not altered the formulation in other ways. Higher potency preparations of Botox®, at 200 units/vial, have also been made available recently, but no similar increase for Dysport® has been marketed.

Table 2.5 Reconstitution stability studies (performed with non-clinical assessments) under various conditions for the different BoNT product families show a range of conditions and diluents tested over the past 20 years

Authors	Year	Title	Product	Reference	Diluent	Summary	Conclusions[a]
Gartlan, M. G. Hoffman, H. T.	1993	Crystalline preparation of botulinum toxin type A (Botox): degradation in potency with storage	Botox®	Otolaryngol. Head Neck Surg. 1993; 108 (2), 135–140	Normal saline	Measurement of mouse potency of product reconstituted, immediately frozen and stored for 2 weeks	…a 69.8 % loss in potency was found when Botox was reconstituted, immediately frozen, and then assayed 2 weeks later ($p < 0.0001$). Statistically significant degradation in potency was seen after refrigerator storage for 12 hours ($p = 0.007$), but not for 6 h ($p = 0.16$)
Jabor, M. A. Kaushik, R. Shayani, P. Ruiz-Razura, A. Smith, B. K. Morimoto, K. W. Cohen, B. E.	2003	Efficacy of reconstituted and stored botulinum toxin type A: an electrophysiologic and visual study in the auricular muscle of the rabbit	Botox®	Plast. Reconstr. Surg. 2003; 111 (7): 2419–2426	Normal saline	Assessment of product either freshly reconstituted or stored frozen after reconstitution up to 12 weeks for potency, duration of action (by duration of nerve conduction effects) and microbial contamination	In conclusion, using the rabbit model, it seems that reconstituted and stored botulinum toxin type A has the same initial potency but the duration of action is affected sometime after 2 weeks of storage. No bacterial contamination was associated with storing unpreserved reconstituted botulinum toxin type A for up to 12 weeks

Table 2.5 (continued)

Authors	Year	Title	Product	Reference	Diluent	Summary	Conclusions[a]
Hexsel, D. M. Alencar de Castro, I. Zechmeister, D. Almeida do Amaral, A.	2004	Re: Hexsel, et al. Multicenter, double-blind study of the efficacy of injections with botulinum toxin type A reconstituted up to six consecutive weeks before application	Botox®	Dermatol. Surg. 2004; 30 (5): 823	Normal saline	Brief report on microbiological analysis of vials stored for 14 & 15 weeks including after fully opening. No contamination detected in any vial	Nevertheless, it is recommended that all the precautions in diluting and storage of botulinum toxin type A be observed, especially after opening the vials
Alam, M. Yoo, S. S. Wrone, D. A. White, L. E. Kim, J. Y.	2006	Sterility assessment of multiple use botulinum A exotoxin vials: a prospective simulation	Botox®	J Am Acad Derm 2006; 55 (2): 272–275	Bacterio-static saline	Reconstitution of vials and then repeat use for up to 7 weeks. Total of 127 vials used. End vials tested microbiologically; no evidence of contamination	Routine refrigerator storage of medication vials containing reconstituted botulinum toxin does not result in microbial contamination of the contents even after serial re-extraction of solution from these vials, and after handling of such vials by multiple personnel. Storage and subsequent reuse of botulinum toxin appears safe for at least 7 weeks after reconstitution

Table 2.5 (continued)

Authors	Year	Title	Product	Reference	Diluent	Summary	Conclusions[a]
Elmas, C. Ayhan, S. Tuncer, S. Erdogan, D. Calguner, E. Basterzi, Y. Gozil, R. Bahcelioglu, M.	2007	Effect of fresh and stored botulinum toxin A on muscle and nerve ultrastructure: an electron microscopic study	Botox®	Ann. Plast. Surg. 2007; 59 (3): 316–322	Normal saline	Study on rabbits injected with either saline, freshly reconstituted product or product reconstituted and stored at 4 °C for 2 weeks into anterior auricular muscles. Muscle and motor nerve examined 5 days or 12 weeks after injection	Alterations in muscle and nerve structures after botulinum toxin injection revealed that there is no significant difference between freshly reconstituted and stored toxin for 2 weeks, at the onset of effect. However, when stored toxin was used atrophic changes in the muscles has started to return earlier or it is less severe than the fresh toxin. . . . However, there is no significant difference between the effects of fresh and stored toxin on nerve ultrastructure at 3 months
Menon, J. Murray, A.	2007	Microbial growth in vials of botulinum toxin following use in clinic	Dysport®	Eye (Lond) 2007; 21 (7): 995–997	Unspecified (probably normal saline)	Microbiological assessment of vials reconstitutes and stored at room temperature for 4 h and after refrigeration for 5 days at 3–5 °C. No microbial contamination detected	This pilot study suggests that if aseptic precautions are followed during the use of botulinum toxin, the contents of the bottle remain sterile despite being exposed to room temperatures for up to 4 h

Normal saline is often referred to as preservative-free saline and bacteriostatic saline as preserved/preservative-containing saline (contains 0.9 % benzyl alcohol)

[a] Conclusions presented as verbatim quotations from publication

Table 2.6 Reconstitution stability studies (including clinical assessments), performed under various conditions for the different BoNT product families, show a range of conditions and diluents tested over the past 15 years

Authors	Year	Title	Product	Reference	Diluent	Summary	Conclusions[a]
Sloop, R. R. Cole, B. A. Escutin, R. O.	1997	Reconstituted botulinum toxin type A does not lose potency in humans if it is refrozen or refrigerated for 2 weeks before use	Botox®	Neurology. 1997; 48(1):249–53	Normal saline	Measure of the decline in extensor digitorum brevis (EDB) M-wave amplitude (percent paralysis) following injection of freshly reconstituted BTX compared with the effect of BTX that was refrozen ($-20°C$) or refrigerated ($+4°C$) for 2 weeks after reconstitution	Essentially no difference in the muscle paralysis resulting from fresh BTX compared with refrozen or refrigerated BTX, and no statistical difference between groups was noted
Alam, M. Dover, J. S. Arndt, K. A	2002	Pain associated with injection of botulinum A exotoxin reconstituted using isotonic sodium chloride with and without preservative: a double-blind, randomized controlled trial	Botox®	Arch. Dermatol. 2002; 138 (4): 510–514	Bacteriostatic saline	Prospective (blinded) and retrospective clinical assessments on pain of injection when using either 0.9 % saline (recommended) or 0.9 % saline with benzyl alcohol preservative. Split-face study	Reconstitution of botulinum toxin with preservative-containing saline can markedly decrease patient discomfort at the time of injection. The difference is statistically and clinically significant

Table 2.6 (continued)

Authors	Year	Title	Product	Reference	Diluent	Summary	Conclusions[a]
Hexsel, D. M. De Almeida, A. T. Rutowitsch, M. De Castro, I. A. Silveira, V. L. Gobatto, D. O. Zechmeister, M. Mazzuco, R. Zechmeister, D.	2003	Multicenter, double-blind study of the efficacy of injections with botulinum toxin type A reconstituted up to six consecutive weeks before application	Botox®	Dermatol. Surg. 2003; 29 (5): 523–529	Normal saline	Double-blind study of 88 volunteers treated for glabellar lines with product reconstituted over 6 weeks	BTX-A may be applied up to 6 weeks after reconstitution without losing its effectiveness. Other factors, which are probably individual, may influence the response to BTX-A injections
Trindade De Almeida, A. R. Kadunc, B. V. Di, Chiacchio N. Neto, D. R.	2003	Foam during reconstitution does not affect the potency of botulinum toxin type A	Botox®	Dermatol. Surg. 2003; 29 (5): 530–531	Normal saline	Clinical assessment of vigorous shaking on product efficacy	Our article demonstrates that when reconstituting Botox, the presence of foam does not affect the potency of the product
van Laborde, S. Dover, J. S. Moore, M. Stewart, B. Arndt, K. A. Alam, M.	2003	Reduction in injection pain with botulinum toxin type B further diluted using saline with preservative: a double-blind, randomized controlled trial		J. Am. Acad. Derm 2003; 48 (6): 875–877	Normal saline and bacterio-static saline	Double-blind, randomized study of 15 volunteers in split-face treatment of upper face dynamic lines to determine if diluent affected pain on injection	Use of preservative-containing saline to further dilute botulinum toxin type B can significantly decrease patient discomfort on injection

Table 2.6 (continued)

Authors	Year	Title	Product	Reference	Diluent	Summary	Conclusions[a]
Kwiat, D. M. Bersani, T. A. Bersani, A.	2004	Increased patient comfort utilizing botulinum toxin type a reconstituted with preserved versus non-preserved saline	Botox®	Opthal. Plas. Recon. Surg. 2004; 20 (3): 186–189	Normal saline and bacterio-static saline	Double-blind, split-face clinical study on 20 volunteers for pain on injection using either diluent	Injection of botulinum toxin reconstituted with preserved saline is less painful than non-preserved saline preparations
Sarifakioglu, N. Sarifakioglu, E.	2005	Evaluating effects of preservative-containing saline solution on pain perception during botulinum toxin type-A injections at different locations: a prospective, single-blinded, randomized controlled trial		Aesth. Plas. Surg. 2005; 29 (2): 113–115	Normal saline and bacterio-static saline	Single-blind randomized clinical study of pain in 93 patients injected in face, neck or axilla using either diluent	The authors conclude that the preservative-containing saline solution significantly decreased pain perception during BTX-A injections ($p = 0.000$)
Thomas, J. P. Siupsinskiene, N.	2006	Frozen versus fresh reconstituted botox for laryngeal dystonia	Botox®	Otolaryngol. Head Neck Surg. 2006; 135 (2), 204–208	(Unspecified)	Open label crossover study of 43 patients with adductor spasmodic dysphonia randomly treated with either freshly reconstituted or previously frozen product (2 occasions for up to 8 weeks)	BTX-A may be safely used after being reconstituted and frozen or refrozen without a significant loss of effectiveness or additional side effects. In our experience, the period of freezing was on 2 occasions for up to 8 weeks

Table 2.6 (continued)

Authors	Year	Title	Product	Reference	Diluent	Summary	Conclusions[a]
Hui, J. I. Lee, W. W.	2007	Efficacy of fresh versus refrigerated botulinum toxin in the treatment of lateral periorbital rhytids	Botox®	Ophthal. Plast. Reconstr. Surg. 2007; 23 (6): 433–438	Bacterio-static saline	Double-blind randomized study on 45 volunteers split-face design using product freshly reconstituted or product stored for 2 weeks at 4 °C after reconstitution	This study demonstrates that 2 weeks of refrigeration does not appear to significantly affect the time of onset or efficacy of botulinum toxin in the treatment of lateral periorbital rhytids
Lizarralde, M. Gutierrez, S. H. Venegas, A.	2007	Clinical efficacy of botulinum toxin type A reconstituted and refrigerated 1 week before its application in external canthus dynamic lines	Botox®	Dermatol. Surg. 2007; 33 (11): 1328–1333	Normal saline	Double-blind, randomized clinical study of 30 volunteers treated in canthal lines with product either freshly reconstituted or stored after reconstitution at 4 °C for 1 week. Assessments by photographs or nerve conduction	BTX-A, reconstituted and refrigerated 1 week before its application, has similar clinical efficacy in treating external canthus dynamic lines as does fresh BTX-A
Parsa, A. A. Lye, K. D. Parsa, F. D.	2007	Reconstituted botulinum type A neurotoxin: clinical efficacy after long-term freezing before use	Botox® Cos-metic	Aesthetic Plast. Surg. 2007; 31(2):188–91: discussion 192–193	Normal saline	Clinical assessment of freezing Botox® in syringes after reconstitution for up to 6 months compared to unfrozen product used within 4 hours	Reconstituted BoNT/A may be frozen, thawed, and injected without losing its potency up to 6 months, with efficacy equivalent to freshly prepared BoNT/A

Table 2.6 (continued)

Authors	Year	Title	Product	Reference	Diluent	Summary	Conclusions[a]
Hexsel, D. Rutowitsch, M. Castro, L. Zechmeister-do-Prado, D.	2008	Multicenter double-blind study of the efficacy of injections with a commercial preparation of botulinum toxin type A reconstituted in different weeks	Dysport®	J. Am. Acad. Dermatol. 2008; 58 (2) suppl. Presented as a poster at AAD meeting, 2008	Normal saline	Clinical assessment of 105 volunteers (3 groups) treated for glabellar lines with product reconstituted for 8 h, 8 days and 15 days before use. No microbial contamination detected	This study showed that the time of dilution up to 15 days before applications had no influence on the efficacy and safety of the 500 U BT-A
Kazim, N. A. Black, E. H.	2008	Botox: shaken, not stirred	Botox®	Ophthal. Plast. Reconstr. Surg. 2008; 24 (1): 10–12	Normal saline	Double-blinded, randomized 6-month trial of seven volunteers injected in frontalis with product reconstituted as manufacturer's instructions or vigorously shaken for 30 s	The effect of botulinum toxin type A is maintained and has the same duration when it is reconstituted vigorously compared with when it is reconstituted gently
Yang, G. C. Chiu, R. J. Gillman, G. S.	2008	Questioning the need to use Botox within 4 hours of reconstitution: a study of fresh vs 2-week-old Botox	Botox®	Arch. Facial. Plast. Surg. 2008; 10 (4): 273–279	Normal saline	Double-blind, randomized trial of 40 volunteers treated for forehead rhytides using product freshly reconstituted or stored for 2 weeks at 4 °C or −20 °C	No measurable difference was found in the potency or duration of efficacy of Botox in the treatment of forehead rhytides after 2 weeks of refrigeration or freezing compared with fresh reconstituted Botox

Table 2.6 (continued)

Authors	Year	Title	Product	Reference	Diluent	Summary	Conclusions[a]
Hexsel, D. Rutowitsch, M. S. De Castro, L. C. Do Prado, D. Z. Lima, M. M.	2009	Blind multicenter study of the efficacy and safety of injections of a commercial preparation of botulinum toxin type A reconstituted up to 15 days before injection	Dysport®	Dermatol. Surg. 2009; 35: 933–940	Normal saline	(Full publication of 2008 poster)	The results confirm the possibility of injecting 500U of BT-A up to 15 days after its reconstitution safely and without loss of efficacy
Shome, D. Nair, A. G. Kapoor, R. Jain, V.	2010	Botulinum toxin A: is it really that fragile a molecule?	Botox®	Dermatol. Surg. 2010; 36 (suppl. 4): 2106–2110	Normal saline	Potency study in mice examining reconstituted product inverted 30 times a minute for up to 6 weeks at 4 °C	OnabotulinumtoxinA is an extremely stable molecule, and vigorous agitation does not impair its potency, even after 6 weeks
Allen, S. B. Goldenberg, N. A.	2012	Pain difference associated with injection of abobotulinumtoxinA reconstituted with preserved saline and preservative-free saline: a prospective, randomized, side-by-side, double-blind study	Dysport®	Derm. Surg. 2012; 38 (6): 867–870	Normal saline and bacterio-static saline	Double-blind, randomized study of 20 volunteers over 2 weeks in split-face treatment of glabellar lines and crow's feet to determine if diluent affected pain on injection	Reconstitution of abobotulinumtoxinA with preserved saline results in significantly less pain on injection than with preservative-free saline. Preserved saline may be the reconstituent of choice for reconstitution of abobotulinumtoxinA

Normal saline is often referred to as preservative-free saline and bacteriostatic saline as preserved/preservative-containing saline (contains 0.9 % benzyl alcohol)

[a]Conclusions presented as verbatim quotations from publication

So where to now with the formulations? What new approaches are being considered or studied?

Pickett and Perrow [90] have discussed new formulations, in particular replacement of human plasma-derived HSA with a fully recombinant version; but this change has not appeared in commercial products to date. The replacement of other components to 'modernise' the formulations was also discussed in their work [90].

Pickett [135] has considered whether a liquid formulation might appear in the future, perhaps in pre-filled syringes or similar administration devices, effectively removing the need for reconstitution.

The concept of a liquid formulation for BoNT products has been in existence for many years. BoNT type B product has already been provided as a liquid in a vial since license approval in December 2000. This product may be painful when injected, likely to be due to the acidic pH of the solution [136]. One key to a successful future liquid product is, therefore, a formulation that is not painful to the patient and which has a physiological solution for injection. Shelf-life stability will also be essential to compete with the existing products, as discussed above.

The patent literature contains numerous references to liquid formulations, filed by Allergan, Ipsen, Elan/Solstice and MedyTox, together with other individuals (Table 2.7). These formulations all have different stabilisers and components, including the omission of HSA. As such, they should be considered next-generation formulations and will be made available as soon as stability and presentation issues are determined by the manufacturers. Also, these new formulations will have to show that they work equivalently in the clinical applications when compared to the current marketed products!

Devices are now starting to appear that are designed to make BoNT injections easier. An example is the botulinum toxin automatic injector called Talent™ BT, recently presented by the Swiss company Primequal [137]. This is effectively a repeat pen injector, containing the reconstituted product in a syringe which 'clicks' each time a trigger is pressed. Does this offer any real advantage? Probably not to the experienced clinician. The repeat 'click' could also be somewhat disconcerting to the patients. More importantly, the cost of the devices is high (20 € each). Is there a need for such high precision of injection volumes? This has not been raised to date as an issue by the clinical community. Nevertheless, the device has recently gained innovation awards.

2.9 Pharmacology of Clinical BoNT Products

In accordance with accepted definitions, pharmacology is the study of what drugs are, how they work and what they do (see, for example, [138]). The pharmacology of products for human therapy must be determined. Licensing of the products without these data is not possible. The study of drug pharmacology is therefore a wide subject, with several subsets such as clinical pharmacology, neuropharmacology, pharmacogenetics, pharmacoepidemiology and toxicology. However, in the case of the BoNT products, the major aspect of their pharmacology is the fact that they

Table 2.7 BoNT liquid formulation patents and patent applications 2008 to present (United States Patent and Trademark Office only)

Patent/Application number	Date granted/published	Title	Abstract	Assigned	Comments
7,211,261	May 1, 2007	Stable liquid formulations of botulinum toxin	Liquid formulations of botulinum toxin stable to storage in liquid form at standard refrigerator temperatures for at least 1–2 years and to storage at higher temperatures for at least 6 months	Moyer, E.; Hirtzer, P. Solstice Neurosciences LLC	Succinate-buffered saline pH range 5.4–5.8. With HSA
8,173,138	May 8, 2012	Stable liquid formulations of botulinum toxin	Liquid formulations of botulinum toxin stable to storage in liquid form at standard refrigerator temperatures for at least 1–2 years and to storage at higher temperatures for at least 6 months	Moyer, E.; Hirtzer, P. Solstice Neurosciences, Inc.	Phosphate, phosphate-citrate and succinate buffers pH 5–6. With HSA. Continuation of US 7,211,261
Patent application 20080050352	*Date published* February 28, 2008	Pharmaceutical composition comprising botulinum, a non-ionic surfactant, sodium chloride and sucrose	Botulinum toxin complex or purified neurotoxin in liquid (or solid) with crystalline agent	Webb, P.; White, M.; Partington, J. Ipsen	Includes buffer, sodium chloride, non-ionic surfactant and disaccharide
20080102090	May 1, 2008	Pharmaceutical composition containing BoNT A2	First application involving type A2 in a liquid with range of benefits compared top type A1	Panjwani, N.; Webb, P.; Pickett, A. Ipsen	Buffer containing surfactant, disaccharide and sodium chloride

Table 2.7 (continued)

Patent/Application number	Date granted/published	Title	Abstract	Assigned	Comments
20090291062	November 26, 2009	Protein formulations and methods of making same	Protein and water formulations of low conductivity and/or low osmolality	Fraunhofer, W.; Bartl; A.; Krause; H-J.; Tschoepe, M.; Kaleta, K.	Includes non-ionic surfactants (polysorbate), non-ionizable excipients (sugar alcohol or polyol) and a method of preparation
20100168023	July 1, 2010	Injectable botulinum toxin formulations	Advantages including reduced antigenicity, reduced tendency for localized diffusion, increased duration of efficacy, faster onset of clinical efficacy, and/or improved stability	Ruegg, C.L.; Stone, H.F.; Waugh, J.M. Revance Therapeutics, Inc.	Positively charged carrier with positively charged backbone including polyamino acid
20100291136	November 18, 2010	Pharmaceutical liquid composition of botulinum toxin with improved stability	A liquid pharmaceutical composition of botulinum toxin which is improved in stability. It comprises botulinum toxin, polysorbate 20, and methionine and optionally isoleucine	Jung, H.H.; Yang, G.H.; Kim, H.W.; Woo, H.D.; Rhee, C.H. Medy-Tox Inc.	Specified ranges of concentrations and pH 5.5–7.0
20100330123	December 30, 2010	Albumin-free botulinum toxin formulations	Botulinum toxin formulations that are stabilized without the use of any proteinaceous excipients	Thompson, S.A.; Ruegg, C.L.; Waugh, J.M.	Non-reducing di/tri-saccharide, non-ionic surfactant, physiologically compatible buffer, different salt compositions

Table 2.7 (continued)

Patent/Application number	Date granted/published	Title	Abstract	Assigned	Comments
20110268765	November 3, 2011	Injectable botulinum toxin formulations	Advantages including reduced antigenicity, a reduced tendency to undergo unwanted localized diffusion following injection, increased duration of clinical efficacy or enhanced potency relative, faster onset of clinical efficacy, and/or improved stability	Ruegg, C.L.; Stone, H.F.; Waugh, J.M. Revance Therapeutics, Inc.	Use of a carrier and a suitable formulation to stabilise against diffusion, reduce antigenicity and reduce diffusion
20120107361	May 3, 2012	Albumin-free botulinum toxin formulations	Botulinum toxin formulations that are stabilized without the use of any proteinaceous excipients	Thompson, S.A.; Ruegg, C.L.; Waugh, J.M.	Non-reducing di/tri-saccharide, non-ionic surfactant, physiologically compatible buffer, different salt compositions

Most patents and patent applications also have equivalent European and World registrations, occasionally with different titles are, quite simply, lethal in minute quantities! Classical pharmacokinetic studies on adsorption, distribution, metabolism and excretion (ADME studies), for BoNT products used clinically, have not been reported, indeed not performed for obvious reasons mostly related to limitations on detection. A recent discussion of the subject by Simpson [139] only dealt with BoNT that had caused intoxication through classical routes and not through clinical application. The entire pharmacology of the BoNT products is therefore skewed as to what can actually be achieved in practice and what data can be obtained using standard techniques that is of value in the assessment of their suitability.

There are several reviews of what has been titled the pharmacology of BoNT, but in reality, these are subsets such as clinical pharmacology or the mode of action in animals [95], [109], [140]–[143].

Coupled to this, of course, is the fact that until now every batch of every product released commercially has been potency tested in animals, a unique situation in the pharmaceutical world. Added to this are the independent release procedures enforced by the regulatory authorities, as described earlier. The safety elements are, therefore, highly enforced and the likelihood of a product being released to the market that is either super- or subpotent is remote.

The pharmacological aspects are also significantly biased by the fact that animals are known to have a different sensitivity to the BoNT products than humans (see, for example, [142], [144]). Mice and rats have different sensitivities to each other [142] and are significantly less sensitive than human adults. These essential facts have been overlooked by many scientists studying the effects of BoNT in animals, most especially when they subsequently make extrapolations to human clinical uses [145]. Studies have been carried out where, if extrapolated to adults, a dose of 5,000 Botox® units or greater would have been administered (see, for example, [146]–[148]). Explanations for these differences in BoNT sensitivities between animals and humans are only now finding explanations, probably in relation to the reduced sensitivity of human BoNT receptors compared to those of other species [33].

No detailed review of the various animal studies and the related pharmacological data obtained is possible in the context of the present work since this would warrant a further chapter. But two aspects are discussed below which are representative of past approaches and future directions in this area, as illustrations of the subject.

2.9.1 Attempts to Relate Animal Data to Safety in Humans

One of the biggest uses of animal studies with BoNT, relating to clinical use, started in the late 1990s, when the first work emerged that attempted to relate the results of animal studies with safety of use in humans. A model was adapted from other uses that claimed to demonstrate useful results when BoNT was included [123]. The original model, called the tail suspension test, was adapted shortly after this initial work to be called the digit abduction scoring (DAS) assay. Basically, mice were injected in one hind limb with different doses of BoNT and then suspended by the tail. Trained observers could then 'measure' the angle that the foot digits were extended to in what is normally described as 'the startle response' of the animal [84]. However, the scale of effect is arbitrary and the result qualitative. Doses as high as 30 Botox® units/kg were employed and, as commented by the workers themselves, 'do not reflect clinical doses' [123]. In addition, the species difference between rodents and man was also acknowledged by the investigators.

Nevertheless, this work was evolved to the point at which claims of safety differences between the BoNT products were being made, based upon these animal pharmacological studies [149]. The investigators revised their earlier comments into using the animal data as representative of human doses ('translational medicine').

When these studies were looked at in detail, the results were clearly not as presented. Exceptionally high comparative doses of products were being used (up to 100 units/kg) [149] and the statistical analysis was flawed [150]. The kinetic analyses, which purported to show different product characteristics, were shown subsequently to be identical, when appropriate differences in the potency units of the products under study were taken into account [150]. There was actually no demonstrable difference in the dose–response of the products tested [150].

2.9.2 Modern Approaches to Pharmacological Studies

Since the studies of the early 2000s, several groups have regrettably repeated the errors when trying to understand the mechanisms of clinical action, the pharmacology, of the BoNT products, as discussed above. But new approaches are slowly emerging that could be of significant value in the future.

The recent work of Hale and colleagues [151] has used the technique of near-infrared whole body analysis to examine the pharmacokinetics of a detoxified BoNT molecule in mice. The images clearly show the kinetics of BoNT diffusion throughout the animals (Fig. 2.4). Intramuscular injections in the tail remained localised over the duration of the study (24 H), but intravenous, intraperitoneal and oral administration showed dissemination throughout the animal. The mutated BoNT molecule had previously been shown to bind to neuronal cells and has been well characterised [152], [153].

This modern approach may herald a new era in ADME studies of BoNT. Future data and publications are awaited with great enthusiasm by the BoNT scientific and clinical communities.

2.10 Conclusions

The manufacture and quality aspects of the various main families of clinical BoNT products are defined and established. Although more than 25 years old, the major products are still finding new clinical applications today and, as always has happened, clinicians are leading the way in developing those new uses.

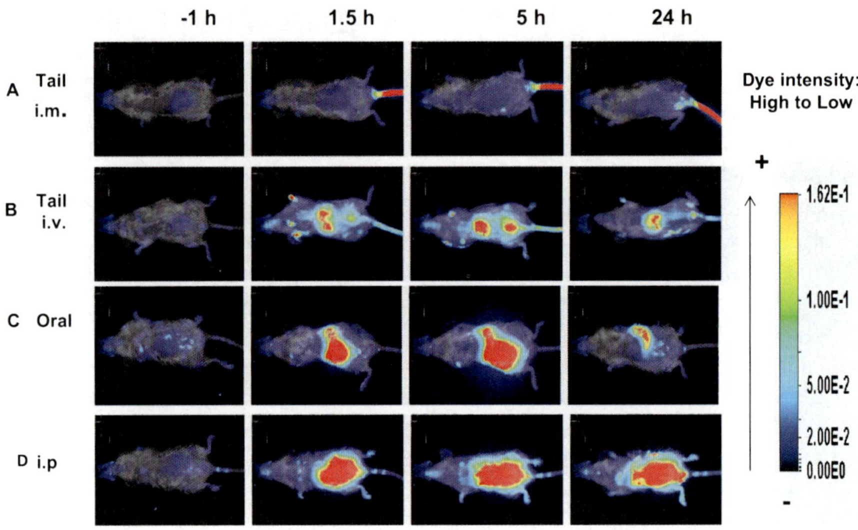

Fig. 2.4 Use of near-infrared technology for whole body in vivo monitoring of metabolism of a detoxified recombinant botulinum toxin labelled with the dye 800 CW. Images for each scan were captured at white light, 700 and 800 nm and fluorescent signals were analysed using the Pearl Cam software supplied with the Odyssey Imaging System (LiCor Inc). (Reprinted from Hale M, Riding S, Sing Bal Ram (2010) Near-infrared imaging of Balb/c mice injected with a detoxified botulinum neurotoxin A. Botulinum J 1(4):431–441, Copyright (2010), with permission)

As the decades pass, new technology on formulations and pharmacological aspects are finding their way into use, to advance even further our understanding of these essential, potent and fascinating products. The coming 25 years are awaited with great anticipation.

References

1. Simpson DM, Blitzer A, Brashear A, Comella C, Dubinsky R, Hallett M, Jankovic J, Karp B, Ludlow CL, Miyasaki JM, Naumann M, So Y (2008) Assessment: botulinum neurotoxin for the treatment of movement disorders (an evidence-based review): report of the therapeutics and technology assessment subcommittee of the American academy of neurology. Neurology 70(19):1699–1706
2. Simpson DM, Gracies JM, Graham HK, Miyasaki JM, Naumann M, Russman B, Simpson LL, So Y (2008) Assessment: botulinum neurotoxin for the treatment of spasticity (an evidence-based review): report of the therapeutics and technology assessment subcommittee of the American academy of neurology. Neurology 70(19):1691–1698
3. Naumann M, So Y, Argoff CE, Childers MK, Dykstra DD, Gronseth GS, Jabbari B, Kaufmann HC, Schurch B, Silberstein SD, Simpson DM (2008) Assessment: botulinum neurotoxin in the treatment of autonomic disorders and pain (an evidence-based review): report of the therapeutics and technology assessment subcommittee of the American academy of neurology. Neurology 70(19):1707–1714

4. Scott AB, Rosenbaum A, Collins CC (1973) Pharmacologic weakening of extraocular muscles. Invest Ophthalmol 12(12):924–927
5. Tsui JK, Eisen A, Mak E, Carruthers J, Scott A, Calne DB (1985) A pilot study on the use of botulinum toxin in spasmodic torticollis. Can J Neurol Sci 12(4):314–316
6. Jankovic J, Brin MF (1991) Therapeutic uses of botulinum toxin. N Engl J Med 324(17): 1186–1194
7. Truong DD, Jost WH (2006) Botulinum toxin: clinical use. Parkinsonism Relat Disord 12(6):331–355
8. Bhidayasiri R, Truong DD (2008) Evidence for effectiveness of botulinum toxin for hyperhidrosis. J Neural Transm 115(4):641–645
9. Truong DD, Bhidayasiri R,(2008) Evidence for the effectiveness of botulinum toxin for sialorrhoea. J Neural Transm 115(4):631–635
10. Watts CR, Truong DD, Nye C (2008) Evidence for the effectiveness of botulinum toxin for spasmodic dysphonia from high-quality research designs. J Neural Transm 115(4):625–630
11. Hexsel C, Hexsel D, Porto MD, Schilling J, Siega C (2011) Botulinum toxin type A for aging face and aesthetic uses. Dermatol Therap 24(1):54–61
12. Jabbari B, Machado D (2011) Treatment of refractory pain with botulinum toxins–an evidence-based review. Pain Med 12(11):1594–1606
13. Linde M, Hagen K, Stovner LJ (2011) Botulinum toxin treatment of secondary headaches and cranial neuralgias: a review of evidence. Acta Neurol Scand Suppl 124(191):50–55
14. Pickett A, Rosales RL (2011) New trends in the science of botulinum toxin-A as applied in dystonia. Int J Neurosci 121(Suppl 1):22–34
15. Gassner HG, Brissett AE, Otley CC, Boahene DK, Boggust AJ, Weaver AL, Sherris DA (2006) Botulinum toxin to improve facial wound healing: A prospective, blinded, placebo-controlled study. Mayo Clin Proc 81(8):1023–1028
16. Chalhoub M, Harris K, Sasso L, Bourjeily G (2011) The use of botox in the treatment of post cardiac surgery paradoxical vocal cord movement. Heart Lung Circ 20(9):602–604
17. Kessler KR, Skutta M, Benecke R (1999) Long-term treatment of cervical dystonia with botulinum toxin A: efficacy, safety, and antibody frequency. J Neurol 246(4):265–274
18. Elovic EP, Brashear A, Kaelin D, Liu J, Millis SR, Barron R, Turkel C (2008) Repeated treatments with botulinum toxin type a produce sustained decreases in the limitations associated with focal upper-limb poststroke spasticity for caregivers and patients. Arch Phys Med Rehabil 89(5):799–806
19. Bentivoglio AR, Fasano A, Ialongo T, Soleti F, Lo FS, Albanese A (2009) Fifteen-year experience in treating blepharospasm with Botox or Dysport: same toxin, two drugs. Neurotox Res 15(3):224–231
20. Pickett A (2011) Consistent biochemical data are essential for comparability of botulinum toxin type A products. Drugs R D 11(1):97–98, (author reply 98–99)
21. Pickett A (2011) Evaluating botulinum toxin products for clinical use requires accurate, complete, and unbiased data. Clin Ophthalmol 5:1287–1290
22. Bentivoglio AR, Ialongo T, Bove F, De Nigris F, Fasano A (2011) Retrospective evaluation of the dose equivalence of Botox ((R)) and Dysport ((R)) in the management of blepharospasm and hemifacial spasm: a novel paradigm for a never ending story. Neurol Sci:1–7
23. Monheit G, Carruthers A, Brandt F, Rand R (2007) A randomized, double-blind, placebo-controlled study of botulinum toxin type A for the treatment of glabellar lines: determination of optimal dose. Dermatol Surg 33(Spec No. 1):51–59
24. Atassi MZ (2004) Basic immunological aspects of botulinum toxin therapy. Mov Disord 19(Suppl 8):68–84
25. Dressler D, Eleopra R (2006) Clinical use of non-A botulinum toxins: botulinum toxin type B. Neurotox Res 9(2–3):121–125
26. Brashear A, Lew MF, Dykstra DD, Comella CL, Factor SA, Rodnitzky RL, Trosch R, Singer C, Brin MF, Murray JJ, Wallace JD, Willmer–Hulme A, Koller M (1999) Safety and efficacy of NeuroBloc (botulinum toxin type B) in type A–responsive cervical dystonia. Neurology 53(7):1439

27. Brin MF, Lew MF, Adler CH, Comella CL, Factor SA, Jankovic J, O'Brien C, Murray JJ, Wallace JD, Willmer-Hulme A, Koller M (1999) Safety and efficacy of NeuroBloc (botulinum toxin type B) in type A-resistant cervical dystonia. Neurology 53(7):1431–1438

28. Lew MF, Adornato BT, Duane DD, Dykstra DD, Factor SA, Massey JM, Brin MF, Jankovic J, Rodnitzky RL, Singer C, Swenson MR, Tarsy D, Murray JJ, Koller M, Wallace JD (1997) Botulinum toxin type B: a double-blind, placebo-controlled, safety and efficacy study in cervical dystonia. Neurology 49(3):701–707

29. Dressler D, Adib Saberi F, Benecke R (2002) Botulinum toxin type B for treatment of axillar hyperhidrosis. J Neurol 249(12):1729–1732

30. Flynn TC, Clark RE 2nd (2003) Botulinum toxin type B (MYOBLOC) versus botulinum toxin type A (BOTOX) frontalis study: rate of onset and radius of diffusion. Dermatol Surg: official publication for American Society for Dermatologic Surgery [et al.] 29(5):519–522

31. Carruthers A, Carruthers J, Flynn TC, Leong MS (2007) Dose-finding, safety, and tolerability study of botulinum toxin type B for the treatment of hyperfunctional glabellar lines. Dermatol Surg: official publication for American Society for Dermatologic Surgery [et al.], 33(1 Spec No.):60–68

32. Oh YJ, Lee NY, Suh DH, Koh JS, Lee SJ, Shin MK (2011) A split-face study using botulinum toxin type B to decrease facial erythema index. J Cosmet Laser Ther 13(5):243–248

33. Strotmeier J, Willjes G, Binz T, Rummel A (2012) Human synaptotagmin-II is not a high affinity receptor for botulinum neurotoxin B and G: increased therapeutic dosage and immunogenicity. FEBS Lett 586(4):310–313

34. Panjwani N, O'Keeffe R, Pickett AM (2008) Biochemical, functional and potency characteristics of type A botulinum toxin in clinical use. Botulinum J 1(1):153–166

35. http://www.edqm.eu/en-Human-Biologicals-OCABR-611.html. Accessed 24 Jan 2011

36. http://www.edqm.eu/en/General-European-OMCL-Network-46.html Accessed 24 Jan 2011

37. Bonventre PF, Kempe LL (1960) Physiology of toxin production by Clostridium botulinum types A and B. I. Growth, autolysis, and toxin production. J Bacteriol 79(1):18–23

38. Rao S, Starr RL, Morris MG, Lin WJ (2007) Variations in expression and release of botulinum neurotoxin in Clostridium botulinum type A strains. Foodborne Pathog Dis 4(2):201–207

39. Day LE, Costilow RN (1964) Physiology of the sporulation process in Clostridium botulinum I. Correlation of morphological changes with catabolic activities, synthesis of dipicolinic acid, and development of heat resistance. J Bacteriol 88:690–694

40. Day LE, Costilow RN (1964) Physiology of the sporulation process in Clostridium botulinum II. Maturation of forespores. J Bacteriol 88:695–701

41. Bradshaw M, Marshall KM, Heap JT, Tepp WH, Minton NP, Johnson EA (2010) Construction of a nontoxigenic Clostridium botulinum strain for food challenge studies. Appl Environ Microbiol 76(2):387–393

42. Cooksley CM, Davis IJ, Winzer K, Chan WC, Peck MW, Minton NP (2010) Regulation of neurotoxin production and sporulation by a putative agrBD signaling system in proteolytic Clostridium botulinum. Appl Environ Microbiol 76(13):4448–4460

43. Bradshaw M, Dineen SS, Maks ND, Johnson EA (2004) Regulation of neurotoxin complex expression in Clostridium botulinum strains 62A, Hall A-hyper, and NCTC 2916. Anaerobe 10(6):321–333

44. Bonventre PF, Kempe LL (1959) Physiology of toxin production by Clostridium botulinum types A and B. II. Effect of carbohydrate source on growth, autolysis, and toxin production. Appl Microbiol 7(6):372–374

45. Bonventre PF, Kempe LL (1959) Physiology of toxin production by Clostridium botulinum types A and B. III. Effect of pH and temperature during incubation on growth, autolysis. and toxin production. Appl Microbiol 7(6):374–377

46. Bonventre PF, Kempe LL (1960) Physiology of toxin production by Clostridium botulinum types A and B. IV. Activation of the toxin. J Bacteriol 79(1):24–32

47. Schantz EJ, Johnson EA (1992) Properties and use of botulinum toxin and other microbial neurotoxins in medicine. Microbiol Rev 56(1):80–99

48. Pickett A, Perrow K (2009) Composition and molecular size of Clostridium botulinum type A toxin-hemagglutinin complex. Protein J 28(5):248–249, (discussion 250–251)
49. Pickett A (2007) Tribute to an enigma: the life and times of Ivan Clifford Hall, 1885–1975, in 44th annual meeting of the Interagency Botulism Research Coordinating Committee. Asilomar, CA
50. Hall IC (1928) A collection of anaerobic bacteria. Science 68(1754):141–142
51. Frevert J (2010) Content of botulinum neurotoxin in botox(r)/vistabel(r), dysport(r)/azzalure(r), and xeomin(r)/bocouture(r). Drugs R D 10(2):67–73
52. Fang PK, Raphael BH, Maslanka SE, Cai S, Singh BR (2010) Analysis of genomic differences among Clostridium botulinum type A1 strains. BMC Genomics 11(1):725
53. Zhang L, Lin WJ, Li S, Aoki KR (2003) Complete DNA sequences of the botulinum neurotoxin complex of Clostridium botulinum type A-Hall (Allergan) strain. Gene 315:21–32
54. Zhang L, Lin W-J, Li S, Aoki KR (2003) Corrigendum to "complete DNA sequences of the botulinum neurotoxin complex of Clostridium botolinum type A-Hall (Allergan) strain" [Gene 315 (2002) 21–32]. Gene 322(0):187
55. Sebaihia M, Peck MW, Minton NP, Thomson NR, Holden MT, Mitchell WJ, Carter AT, Bentley SD, Mason DR, Crossman L, Paul CJ, Ivens A, Wells-Bennik MH, Davis IJ, Cerdeno-Tarraga AM, Churcher C, Quail MA, Chillingworth T, Feltwell T, Fraser A, Goodhead I, Hance Z, Jagels K, Larke N, Maddison M, Moule S, Mungall K, Norbertczak H, Rabbinowitsch E, Sanders M, Simmonds M, White B, Whithead S, Parkhill J (2007) Genome sequence of a proteolytic (group I) Clostridium botulinum strain Hall A and comparative analysis of the clostridial genomes. Genome Res 17(7):1082–1092
56. http://www.kobio.org/kobio/fileupload/downloadFile.bio?fnum=3&seq=1 meditoxin. Accessed 23 Jan 2012
57. Zhao YJ, Wei R, Liu C (2010) Cloning and sequencing of complete BoNT gene of Clostridium botulinum type A for therapy. Chin J Biol 23(6):598–601
58. Siegel LS, Metzger JF (1980) Effect of fermentation conditions on toxin production by Clostridium botulinum type B. Appl Environ Microbiol 40(6):1023–1026
59. Setler P (2000) The biochemistry of botulinum toxin type B. Neurology 55(12 Suppl 5):22–28
60. Eleopra R, Tugnoli V, Rossetto O, Montecucco C, De Grandis D (1997) Botulinum neurotoxin serotype C: a novel effective botulinum toxin therapy in human. Neurosci Lett 224(2):91–94
61. Eleopra R, Tugnoli V, Rossetto O, De Grandis D, Montecucco C (1998) Different time courses of recovery after poisoning with botulinum neurotoxin serotypes A and E in humans. Neurosci Lett 256(3):135–138
62. Eleopra R, Tugnoli V, Quatrale R, Rossetto O, Montecucco C (2002) Different types of botulinum toxin in humans. Naunyn Schmiedebergs Arch Pharmacol 365(Suppl 2):R18
63. Eleopra R, Tugnoli V, Quatrale R, Rossetto O, Montecucco C (2004) Different types of botulinum toxin in humans. Mov Disord 19(Suppl 8):S53–S59
64. Eleopra R, Tugnoli V, Quatrale R, Rossetto O, Montecucco C, Dressler D (2006) Clinical use of non-A botulinum toxins: botulinum toxin type C and botulinum toxin type F. Neurotox Res 9(2–3):127–131
65. Lautenschlager M, Maslanka SE, Paul PA, Kalb SR, Barr JR, Raphael BH (2012) Recovery and detection of botulinum neurotoxins from a nonporous surface. J Microbiol Methods 92(3):278–280
66. Munro K, Lanser J, Flower R (1999) A comparative study of methods to validate formaldehyde decontamination of biological safety cabinets. Appl Environ Microbiol 65(2):873–876
67. Johnston MD, Lawson S, Otter JA (2005) Evaluation of hydrogen peroxide vapour as a method for the decontamination of surfaces contaminated with Clostridium botulinum spores. J Microbiol Methods 60(3):403–411
68. World Health Organisation (2003) Laboratory biosafety manual. Geneva, Switzerland. 99
69. Centers for Disease Control and Prevention, National Institutes of Health (2009) Biosafety in microbiological and biomedical laboratories. US Department of Health and Human Services, Public Health Service, p. 415

70. Advisory Committee on Dangerous Pathogens (2001) The management, design and operation of microbiological containment laboratories. HSE Books, Suffolk, p. 67
71. http://www.bt.cdc.gov/agent/agentlist-category.asp. Accessed 17 March 2011
72. http://www.selectagents.gov. Accessed 17 March 2011
73. Select Agents and Toxins Security Information Document. Prepared by US Department of Health and Human Services (HHS) Center for Disease Control and Prevention (CDC) and US Department of Agriculture Animal and Plant Health Inspection Service (APHIS), March 8 2007
74. Health and Safety Executive (HSE), Advisory Committee on Dangerous Pathogens (ACDP), (2004) The approved list of biological agents, HMSO. United Kingdom
75. Centers for Disease Control and Prevention (2011) Notice of CDC's discontinuation of investigational pentavalent (ABCDE) botulinum toxoid vaccine for workers at risk for occupational exposure to botulinum toxins. MMWR 60(42):1454–1455
76. Henkel JS, Tepp WH, Przedpelski A, Fritz RB, Johnson EA, Barbieri JT (2011) Subunit vaccine efficacy against botulinum neurotoxin subtypes. Vaccine 29(44):7688–7695
77. White DM, Pellett S, Jensen MA, Tepp WH, Johnson EA, Arnason BG (2011) Rapid immune responses to a botulinum neurotoxin Hc subunit vaccine through in vivo targeting to antigen-presenting cells. Infect Immun 79(8):3388–3396
78. Abrams A, Kegeles G, Hottle GA (1946) The purification of toxin from Clostridium botulinum type A. J Biol Chem 164:63–79
79. Lamanna C, McElroy OE, Eklund HW (1946) The purification and crystallization of Clostridium botulinum type A toxin. Science 103(2681):613–614
80. Hambleton P, Capel B, Bailey N et al (1981) Production, purification and toxoiding of Clostridium botulinum type A toxin. In: Lewis Jr GE, Angel PS (ed) Biomedical aspects of botulism, Academic Press Inc, New York
81. Hambleton P (1992) Clostridium botulinum toxins: a general review of involvement in disease, structure, mode of action and preparation for clinical use. J Neurol 239(1):16–20
82. Duff JT, Wright GG, Klerer J, Moore DE, Bibler RH (1957) Studies on immunity to toxins of Clostridium botulinum. I. A simplified procedure for isolation of type A toxin. J Bacteriol 73(1):42–47
83. Schantz EJ, Johnson EA (1997) Botulinum toxin: the story of its development for the treatment of human disease. Perspect Biol Med 40(3):317–327
84. Aoki KR (1999) Preclinical update on BOTOX® (botulinum toxin type A)-purified neurotoxin complex relative to other botulinurn neurotoxin preparations. Eur J Neurol 6:s3–s10
85. Lietzow MA, Gielow ET, Le D, Zhang J, Verhagen MF (2008) Subunit stoichiometry of the Clostridium botulinum type A neurotoxin complex determined using denaturing capillary electrophoresis. Protein J 27(7–8):420–425
86. Lietzow MA, Gielow ET, Le D, Zhang J, Verhagen MF (2008) Composition and molecular size of Clostridium botulinum type A toxin–hemagglutinin complex. Protein J 28(5):250–251
87. FDA CBER Establishment Inspetion Report, Allergan, Inc, October 13-20 1997: redacted
88. Wortzman MS, Pickett A (2009) The science and manufacturing behind botulinum neurotoxin type A-ABO in clinical use. Aesthet Surg J 29(6):S34–S42
89. Pickett A, Caird D (2008) Comparison of type a botulinum toxin products in clinical use. J Clin Pharm Ther 33(3):327–328
90. Pickett A, Perrow K (2010) Formulation composition of botulinum toxins in clinical use. J Drugs Dermatol 9(9):1085–1091
91. Pickett A (2013) Reviews of botulinum toxin products in aesthetic use must be accurate, clear and avoid speculation. Clin Pharmacol: Advances and Applications 5:149–152
92. Lorenc ZP, Kenkel JM, Fagien S, Hirmand H, Nestor MS, Sclafani AP, Sykes JM, Waldorf HA (2013) IncobotulinumtoxinA (Xeomin): background, mechanism of action, and manufacturing. Aesthet Surg J 33(Suppl 1):18S–22S
93. Bigalke H, Frevert J (2011). Therapeutic composition comprising a botulinum neurotoxin. US patent 7,964,199 B1 June 21 2011

94. Callaway JE, Arezzo JC, Grethlein AJ (2001) Botulinum toxin type B: an overview of its biochemistry and preclinical pharmacology. Semin Cutan Med Surg 20(2):127–136
95. Callaway JE (2004) Botulinum toxin type B (Myobloc): pharmacology and biochemistry. Clin Dermatol 22(1):23–28
96. Ohishi I, Sakaguchi G (1977) Activation of botulinum toxins in the absence of nicking. Infect Immun 17(2):402–407
97. Evans DM, Williams RS, Shone CC, Hambleton P, Melling J, Dolly JO (1986) Botulinum neurotoxin type B. Its purification, radioiodination and interaction with rat-brain synaptosomal membranes. Eur J biochem/FEBS 154(2):409–416
98. Goodnough MC, Johnson EA (1992) Stabilization of botulinum toxin type A during lyophilization. Appl Environ Microbiol 58(10):3426–3428
99. Elston JS (1990) Botulinum toxin A in clinical medicine. J Physiol (Paris) 84(4):285–289
100. Hambleton P, Cohen HE, Palmer BJ, Melling J (1992) Antitoxins and botulinum toxin treatment. BMJ 304(6832):959–960
101. Zuber M, Sebald M, Bathien N, de Recondo J, Rondot P (1993) Botulinum antibodies in dystonic patients treated with type A botulinum toxin: frequency and significance. Neurology 43(9):1715–1718
102. Greene P, Fahn S, Diamond B (1994) Development of resistance to botulinum toxin type A in patients with torticollis. Mov Disord 9(2):213–217
103. Botox prescribing information, revised October 2006, Allergan Inc, CA
104. Botox prescribing information, October 2010, Allergan Inc, CA reference numbers 71580US11B & 72284US12B
105. Naumann M, Carruthers A, Carruthers J, Aurora SK, Zafonte R, Abu-Shakra S, Boodhoo T, Miller-Messana MA, Demos G, James L, Beddingfield F, VanDenburgh A, Chapman MA, Brin MF (2010) Meta-analysis of neutralizing antibody conversion with onabotulinumtoxinA (BOTOX(R)) across multiple indications. Mov Disord 25(13):2211–2218
106. Racette BA, McGee-Minnich L, Perlmutter JS (1999) Efficacy and safety of a new bulk toxin of botulinum toxin in cervical dystonia: a blinded evaluation. Clin Neuropharmacol 22(6):337–339
107. Jankovic J, Vuong KD, Ahsan J (2003) Comparison of efficacy and immunogenicity of original versus current botulinum toxin in cervical dystonia. Neurology 60(7):1186–1188
108. Naumann M, Yakovleff A, Durif F, B C D P S Group (2002) A randomized, double-masked, crossover comparison of the efficacy and safety of botulinum toxin type A produced from the original bulk toxin source and current bulk toxin source for the treatment of cervical dystonia. J Neurol 249(1):57–63
109. Aoki KR (2005) Pharmacology and immunology of botulinum neurotoxins. Int Ophthalmol Clin 45(3):25–37
110. Atassi MZ, Dolimbek BZ, Steward LE, Aoki KR (2007) Molecular bases of protective immune responses against botulinum neurotoxin A–how antitoxin antibodies block its action. Crit Rev Immunol 27(4):319–341
111. Dolimbek BZ, Aoki KR, Steward LE, Jankovic J, Atassi MZ (2007) Mapping of the regions on the heavy chain of botulinum neurotoxin A (BoNT/A) recognized by antibodies of cervical dystonia patients with immunoresistance to BoNT/A. Mol Immunol 44(5):1029–1041
112. Lange O, Bigalke H, Dengler R, Wegner F, deGroot M, Wohlfarth K (2009) Neutralizing antibodies and secondary therapy failure after treatment with botulinum toxin type A: much ado about nothing? Clin Neuropharmacol 32(4):213–218
113. Pickett A, Panjwani N, O'Keeffe RS (2003) Potency of type A botulinum toxin preparations in clinical use, in 40th annual meeting of the Interagency Botulinum Research Coordinating Committee. Atlanta, USA
114. Dressler D (2004) Clinical presentation and management of antibody-induced failure of botulinum toxin therapy. Mov Disord 19(Suppl 8):S92–S100
115. Stell R, Coleman R, Thompson P, Marsden CD (1988) Botulinum toxin treatment of spasmodic torticollis. BMJ 297(6648):616

116. Racette BA, Stambuk M, Perlmutter JS (2002) Secondary nonresponsiveness to new bulk botulinum toxin A (BCB2024). Mov Disord 17(5):1098–1100
117. Panjwani N, Pickett A, O'Keeffe R (2005) Botulinum type A toxins in clinical use: a comparison of specific potency and protein content. Neurotox Res 9(2–3):225–254
118. Pickett A, O'Keeffe R, Panjwani N (2007) The protein load of therapeutic botulinum toxins. Eur J Neurol 14(4):e11
119. Dodd SL, Rowell BA, Vrabas IS, Arrowsmith RJ, Weatherill PJ (1998) A comparison of the spread of three formulations of botulinum neurotoxin A as determined by effects on muscle function. Eur J Neurol 5(2):181–186
120. Dodd SL, Selsby J, Payne A, Judge A, Dott C (2005) Botulinum neurotoxin type A causes shifts in myosin heavy chain composition in muscle. Toxicon 46(2):196–203
121. Pickett A, O'Keeffe R, Judge A, Dodd S (2008) The in vivo rat muscle force model is a reliable and clinically relevant test of consistency among botulinum toxin preparations. Toxicon 52(3):455–464
122. Cichon JV Jr, McCaffrey TV, Litchy WJ, Knops JL (1995) The effect of botulinum toxin type A injection on compound muscle action potential in an in vivo rat model. Laryngoscope 105(2):144–148
123. Aoki KR, Peng K, Siddiqui T, Spanoyannis A (1995) Pharmacology of BOTOX (botulinum toxin type A) purified neurotoxin complex: local versus systemic muscle activity measurements in mice. Eur J Neurol 2(Suppl 3):3–9
124. http://www.accessdata.fda.gov/drugsatfda_docs/nda/2009/125724Orig1s001MedR.pdf. Accessed 22 Jan 2012
125. Karst KR (2004) Presages to the coming war over generic biologics. J Generic Med 1(2): 155–163
126. Jost WH, Blumel J, Grafe S (2007) Botulinum neurotoxin type A free of complexing proteins (XEOMIN) in focal dystonia. Drugs 67(5):669–683
127. Quarta M (2009) Bilateral comparison of two botulinum toxin type A solutions for lateroorbital facial lines. Double blind study of Botox©versus Xeomin©. Journal für Ästhetische Chirurgie 2(4):199–202
128. Bowe C (2011) Biologic drugs shelf life dating periods: how long can you go? Scrip Intelligence
129. Grein S, Fink K (2009) Complexing proteins are not required for stability of botulinum neurotoxin type A preparations. J Neurol Sci 285(S1):S299
130. Grein S, Mander G, Fink K (2011) Stability of botulinum neurotoxin type A, devoid of complexing proteins. Botulinum J 2(1):49–57
131. Matejtschuk P, Dash CH, Gascoigne EW (2000) Production of human albumin solution: a continually developing colloid. Br J Anaesth 85(6):887–895
132. Hexsel D, Rutowitsch MS, De Castro LC, Do Prado DZ, Lima MM (2009) Blind multicenter study of the efficacy and safety of injections of a commercial preparation of botulinum toxin type A reconstituted up to 15 days before injection. Dermatol Surg 35:933–940
133. Yang GC, Chiu RJ, Gillman GS (2008) Questioning the need to use botox within 4 hours of reconstitution: a study of fresh vs 2-week-old Botox. Arch Facial Plast Surg 10(4):273–279
134. Cote TR, Mohan AK, Polder JA, Walton MK, Braun MM (2005) Botulinum toxin type A injections: adverse events reported to the US food and drug administration in therapeutic and cosmetic cases. J Am Acad Dermatol 53(3):407–415
135. Pickett A (2011) Researcher predicts future generations of neurotoxins. Aesth Guide July/August:36–44
136. Myobloc prescribing information at http://www.myobloc.com/hp_about/PI_5-19-10.pdf. Accessed 22 Jan 2012
137. http://www.primequal.com/home.php. Accessed 22 Jan 2012
138. http://medical-dictionary.thefreedictionary.com/pharmacology. Accessed 22 January 2012
139. Simpson L (2013) The life history of a botulinum toxin molecule. Toxicon 68:40–59
140. Aoki KR (2003) Pharmacology and immunology of botulinum toxin type A. Clin Dermatol 21(6):476–480

141. Aoki KR (2004) Pharmacology of botulinum neurotoxins. Operative techniques in otolaryngology-head and neck surgery 15(2):81–85
142. Rosales RL, Bigalke H, Dressler D (2006) Pharmacology of botulinum toxin: differences between type A preparations. Eur J Neurol 13(Suppl 1):2–10
143. Dressler D, Benecke R (2007) Pharmacology of therapeutic botulinum toxin preparations. Disabil Rehabil 29(23):1761–1768
144. Poulain B, Popoff MR, Molgo J (2008) How do the botulinum neurotoxins block neurotransmitter release: from botulism to the molecular mechanism of action. Botulinum J 1(1):14–87
145. Pickett A (2012) Animal studies with botulinum toxins may produce misleading results. Anesth Analg 115(3):736, author reply 736–7
146. Clowry GJ, Walker L, Davies P (2006) The effects of botulinum neurotoxin A induced muscle paresis during a critical period upon muscle and spinal cord development in the rat. Exp Neurol 202(2):456–469
147. Choi WH, Song CW, Kim YB, Ha CS, Yang GH, Woo HD, Jung HH, Koh WS (2007) Skeletal muscle atrophy induced by intramuscular repeated dose of botulinum toxin type A in rats. Drug Chem Toxicol 30(3):217–227
148. Favre-Guilmard C, Auguet M, Chabrier PE (2009) Different antinociceptive effects of botulinum toxin type A in inflammatory and peripheral polyneuropathic rat models. Eur J Pharmacol 617(1–3):48–53
149. Aoki KR, Ranoux D, Wissel J (2006) Using translational medicine to understand clinical differences between botulinum toxin formulations. Eur J Neurol: the official journal of the European Federation of Neurological Societies 13(Suppl 4):10–19
150. Pickett A (2009) Dysport: pharmacological properties and factors that influence toxin action. Toxicon 54(5):683–689
151. Hale M, Riding S, Singh BR (2010) Near-infrared imaging of balb/c mice injected with a detoxified botulinum neurotoxin A. Botulinum J 1(4):431–441
152. Yang W, Lindo P, Riding S, Chang T-W, Cai S, Van T, Kukreja R, Zhou Y, Vasa K, Singh BR (2008) Expression, purification and comparative characterisation of enzymatically deactivated recombinant botulinum neurotoxin type A. Botulinum J 1(2):219–241
153. Baskaran P, Lehmann TE, Topchiy E, Thirunavukkarasu N, Cai S, Singh BR, Deshpande S, Thyagarajan B (2013) Effects of enzymatically inactive recombinant botulinum neurotoxin type A at the mouse neuromuscular junctions. Toxicon 72:71–80

Chapter 3
Clinical Use of Botulinum Neurotoxin: Neuromuscular Disorders

Arianna Guidubaldi, Anna Rita Bentivoglio and Alberto Albanese

Abstract The original clinical application of botulinum neurotoxin was in the treatment of strabismus by local chemical denervation of the neuromuscular junction and relaxation of the muscle with a duration of several months. This initial application has been followed by use of the neurotoxin to treat a wide range of disorders of muscle hyper-contraction. Botulinum neurotoxin is now a major clinical product for the treatment of spasticity and muscle hyperactivity. Muscle relaxation also underpins the cosmetic use of the neurotoxin. This chapter will review and assess the clinical utility of the various botulinum products in neuromuscular disorders.

Keywords Botulinum neurotoxin · Dystonia · Blepharospasm · Spasticity · Tremor · Dyskinesias

3.1 Introduction

In the late 1970s, a botulinum neurotoxin (BoNT) was introduced as a therapeutic agent for the treatment of strabismus [1]. This pioneering indication has paved the way to the use of BoNT products as therapeutic agents for a wide range of disorders with muscle hyper-contraction. The list of potential applications of BoNT in clinical practice has rapidly expanded to encompass dystonia syndromes, tremor, tics, spasticity and other neuromuscular disorders (Table 3.1). We review here this wealth of information and highlight the therapeutic role of BoNT in neuromuscular disorders.

Several BoNT preparations are now licensed for clinical use [2]. Three branded products contain BoNT/A (onabotulinumtoxinA marketed as Botox®, abobotulinumtoxinA marketed as Dysport®, incobotulinumtoxinA marketed as Xeomin®) and one contains BoNT/B (rimabotulinumtoxinB marketed as Myobloc® in Canada, the USA and Korea and as NeuroBloc® in the European Union, Norway and Iceland). These products are dosed using noninterchangeable proprietary units and switching from

A. Albanese (✉) · A. Guidubaldi · A. R. Bentivoglio
Istituto di Neurologia, Università Cattolica del Sacro Cuore,
Fondazione IRCCS Istituto Neurologico Carlo Besta, Milan, Italy
e-mail: alberto.albanese@unicatt.it

K. A. Foster (ed.), *Clinical Applications of Botulinum Neurotoxin,*
Current Topics in Neurotoxicity 5, DOI 10.1007/978-1-4939-0261-3_3,
© Springer Science+Business Media New York 2014

Table 3.1 BoNT indications in neuromuscular disorders: first introduction

Year	Disease	First report
1980	Strabismus	[1]
1985	Blepharospasm	[24]
1985	Cervical dystonia	[57]
1986	Hemifacial spasm	[144]
1986	Spasmodic dysphonia	[240]
1989	Oromandibular dystonia	[88]
1989	Focal hand dystonia	[241]
1989	Spasticity	[242]
1992	Cosmetic use	[243]
1990	Tardive dyskinesias	[212]
1993	Cerebral palsy in children	[181]
1994	Dystonic tics	[202]
1994	Axial dystonia	[244]
1995	Focal lower limb dystonia	[127]
1997	Freezing of gait	[236]

one to another requires expert clinical management. Licensing varies among products and between countries, particularly within Europe, and expands continuously on indications (Table 3.2).

There is no consensus on how to perform BoNT injections in different neuromuscular disorders. Variables such as dosing, dilutions, number of injections per site, targeting (visual, electromyography (EMG)- or ultrasound-guided) influence outcome and reduce comparability of data among different centers. BoNT injections can be intramuscular, subcutaneous, intradermic or intraglandular and are part of a comprehensive treatment plan.

3.2 Neurological Indications

3.2.1 Dystonia

Dystonia is characterized by sustained muscle contractions, frequently causing repetitive twisting movements or abnormal postures resulting in a combination of dystonic movements and postures [3]. This was the first hyperkinetic movement disorder treated with BoNT [4]. Localized injections provide a transient symptomatic relief in primary and non-primary dystonia syndromes, as demonstrated by several randomized controlled studies and by a large number of uncontrolled studies. Experience on the use of BoNT treatment, in focal dystonias, dates back to almost 30 years ago. Due to this long-lasting experience, treatment of dystonia is currently standardized across movement disorder clinics.

BoNT is the first-choice treatment for most types of focal dystonia. It is established that BoNT/A products, in properly adjusted doses, are effective and safe treatments of primary cranial (excluding oromandibular) and cervical dystonia and are effective on writing dystonia [5]. RimabotulinumtoxinB is also an efficacious treatment for cervical dystonia, but the larger doses required (compared to BoNT/A), pain at

Table 3.2 Approved indications for BoNT use in neuromuscular disorders

Product name/Toxin type	US approved uses	EU approved uses
OnabotulinumtoxinA	Cervical dystonia	Cervical dystonia
	Blepharospasm	Blepharospasm
	Hemifacial spasm	Hemifacial spasm
	Strabismus	Strabismus
	Upper limb spasticity	Focal spasticity
	Glabellar rhytides	Cerebral palsy
		Glabellar rhytides
AbobotulinumtoxinA	Cervical dystonia	Cervical dystonia
	Neck pain	Blepharospasm
	Glabellar rhytides	Hemifacial spasm
		Hyperhidrosis
		Strabismus
		Focal spasticity
		Cerebral palsy
		Glabellar rhytides
IncobotulinumtoxinA	Cervical dystonia	Cervical dystonia
	Blepharospasm	Blepharospasm
		Upper limb spasticity
		Glabellar rhytides
RimabotulinumtoxinB	Cervical dystonia	Cervical dystonia

injection sites and shorter duration of action make it a second-choice option in treating dystonia [6], [7]. BoNT/A is also effective for focal upper limb and laryngeal dystonia, but the results are not as convincing as those collected in cases of cranial and cervical dystonia [8]. The level of evidence for efficacy on focal lower limb dystonia is even lower [8].

Given the long-standing experience in performing treatments, in recent years several long-term studies on the efficacy and safety of onabotulinumtoxinA and abobotulinumtoxinA have been published confirming their safety and efficacy [9]–[11]. IncobotulinumtoxinA has been introduced in Europe and North America only recently and long-term data on this product are not available. Few studies on rimabotulinumtoxinB have been performed in cervical dystonia and blepharospasm, and very few ones in oromandibular and upper limb dystonia. The results of these studies have confirmed the efficacy of rimabotulinumtoxinB but have not responded to the concern about antigenicity and systemic anticholinergic adverse effects. Shared experience on rimabotulinumtoxinB is insufficient compared to the large amount of information published on BoNT/A serotypes [12], [13].

There is informal agreement, albeit no consensus, on the practicalities of BoNT injections for dystonia. Overactive muscles can be identified by direct inspection or by EMG-guided targeting. As mentioned above, direct inspection is usually sufficient to target a superficial muscle, such as most facial and some cervical muscles. In these regions, EMG- or, less commonly, ultrasound-guided targeting provides a second-line approach whenever improvement of muscle selection is needed. In other body regions, such as the sublingual muscles, larynx and limbs, targeting is performed using EMG guidance rather than inspection.

3.2.2 *Blepharospasm*

Blepharospasm is a focal dystonia involving the orbicularis oculi and periocular muscles; when associated with oromandibular involvement, it is referred to as "Meige syndrome." In blepharospasm, there is excessive (intermittent or persistent) involuntary closure of the eyelids, usually bilateral, though it may sometimes be unilateral at onset. Eye closure is produced by phasic or tonic contractions of the ocular muscles. Over time, these may become more frequent and continuous, leading to sustained eyelid closure and functional blindness [14]. Blepharospasm typically begins insidiously between the fifth and the seventh decades. The estimated prevalence increases with age, ranging from as little as 16 cases per million to as many as 133 per million [15], suggesting that in many cases blepharospasm remains underdiagnosed. It affects twice as many women as men [16].

Initially, blinking may increase in response to bright light, accompanied by a sensation of eye discomfort. Symptoms then progress very slowly and the eyes may involuntary shut for long intervals, interrupting the patient's daily activities, such as driving or reading. In its most severe form, blepharospasm results in depression and social isolation. Spasms are absent during sleep. The condition generally takes several years to worsen and it may progress very mildly in some patients. Spontaneous remission occurs rarely, most often within the first 5 years [17]. Patients who develop blepharospasm may experience spread of the dystonia to other body parts. In a recent update on blepharospasm, studies that evaluated spread to other body regions were reviewed [18]. A series of 602 patients with primary dystonia showed that in patients with blepharospasm, spread of the dystonia to other body parts was more likely than in those with others focal forms [19]. Most spread occurred during the first 2 years after onset of blepharospasm, whereas the risk of spread remained roughly constant over time for other dystonias. This is in keeping with other observations that the time from onset to initial spread is shorter in patients with blepharospasm [20], [21].

In the majority of cases, no identifiable cause of blepharospasm is found, and secondary cases account for only 10 % of patients [22]. Therefore, primary (essential) blepharospasm is cured symptomatically. A commonly shared hypothesis is that blepharospasm is related to hyperexcitability of brainstem neurons, as a result of basal ganglia dysfunction. Recently, it has been proposed that an abnormal corneal input induced by excessive blinking may exacerbate increased long-term potentiation type of plasticity, thus leading to blepharospasm [23]. Secondary blepharospasm can occur in response to provocative, irritating mechanical or light stimuli, commonly because of a number of ocular disorders, such as blepharitis, trichiasis, dry eye syndrome and corneal disorders. Additionally, blepharospasm can be observed in a variety of neurodegenerative disorders.

BoNT/A has been quickly recognized as the treatment of choice for blepharospasm (14). Prior to its innovative introduction [24], medical and surgical treatments were rarely successful. Although there is no high-quality, randomized, controlled efficacy data to support the use of BoNT in blepharospasm, several open-label studies on large series indicate that it is an efficacious and safe treatment [25]. The Food and

Drug Administration (FDA) approved BoNT for the treatment of blepharospasm in 1989. Its efficacy has been confirmed by more than 50 open-label studies (accounting for > 2,500 patients), and by a few controlled studies. Data compiled by American Society of Ophthalmology showed that BoNT/A successfully treats approximately 90 % of blepharospasm patients [26]. In keeping with this, guidelines produced by the American Academy of Neurology and the European Federation of Neurological Societies provide class A recommendation that BoNT/A (or BoNT/B if there is immunoresistance to serotype A) is a first-line treatment for primary cranial (excluding oromandibular) or cervical dystonia [5], [8]. Other studies have also evidenced improvement in quality of life after BoNT treatment [27].

Injections are typically well tolerated, with dry eye, eyelid ptosis, and mild facial weakness reported as the most frequent adverse events [26], occurring in less than 10 % of treated patients, and normally of short duration (less than 2 weeks). Treatment commonly starts with small doses that are increased as needed at successive treatment sessions. The upper limit is found when motor improvement lasting for 2–3 months without appreciable side effects is obtained. The site of injection greatly influences the outcome. Best results are obtained when low doses of BoNT are placed at pretarsal, rather than orbital, sites [28]–[31].

Common doses are 20–40 onabotulinumtoxinA U, 75–175 abobotulinumtoxinA U or 2,500 rimabotulinumtoxinB U. Higher doses are reported in some publications, indicating that the therapeutic window for BoNT may be quite wide. Very high doses of onabotulinumtoxinA (> 100 U) have been used in selected cases to treat refractory blepharospasm [32], [33]. The average latency from the time of injection to the onset of improvement varies from 3 to 5 days; a benefit lasting for 2–3 months is observed in almost all patients. The effect of BoNT/A is reversible and in most cases injections are repeated approximately every 3–4 months. Common reasons for lack of efficacy include underdosing and improper injection technique (particularly placement). Secondary resistance to BoNT is rare and can often be managed [34]; the doses used are lower than in other dystonia types and injections are performed less frequently.

Several studies have compared different formulations of BoNT in the treatment of blepharospasm. No differences were found between onabotulinumtoxinA and abobotulinumtoxinA with regard to duration of effect and adverse events in a single-blind, randomized comparison [35]. Based on these data, a 4:1 conversion rate was suggested for blepharospasm and hemifacial spasm (HFS). A double-blind, crossover study on 212 subjects compared abobotulinumtoxinA and onabotulinumtoxinA, using the same 4:1 ratio [36]. The duration of effect was identical in the two groups, but onabotulinumtoxinA caused fewer side effects, particularly ptosis. Another study reported different results about the duration of these two BoNT/A brands. A class IV trial found that onabotulinumtoxinA is more efficacious than abobotulinumtoxinA in blepharospasm and has a longer duration of effect [37]. More recently, we compared a large series of patients with blepharospasm who had been treated with abobotulinumtoxinA or onabotulinumtoxinA for more than 15 years [9]. In this long follow-up, abobotulinumtoxinA had a longer duration of improvement and produced more side effects than onabotulinumtoxinA. Both BoNT/A brands were effective and safe in patients with blepharospasm but had marked differences related to patient management. It has been anecdotally reported that abobotulinumtoxinA is potentially effective in secondary nonresponders to onabotulinumtoxinA, but this has not been confirmed by controlled trials [38].

IncobotulinumtoxinA, which has been licensed recently, has been reported to be not inferior to onabotulinumtoxinA for the treatment of blepharospasm. IncobotulinumtoxinA was compared to onabotulinumtoxinA in a 1:1 dose ratio in two randomized, double-blind paralleled studies and no inferiority of efficacy or difference in tolerability was found [39], [40]. Another double-blind, parallel-group, multicenter study also reported that incobotulinumtoxinA is effective in blepharospasm and does not differ from onabotulinumtoxinA in terms of potency, duration or adverse reaction profile [41].

Another BoNT/A brand (called Prosigne®) is available in China and few other countries. This has not been widely investigated, and few data are available. In a small crossover study on 8 patients with blepharospasm, this toxin brand provided equivalent improvement, with latency, duration and side effects similar to onabotulinumtoxinA [42]. A prospective, randomized, double-blind study has compared this product to onabotulinumtoxinA in blepharospasm and HFS. The mean duration of efficacy was comparable (11.3 weeks for either toxin in blepharospasm). Pain and burning during the injection and the result of the treatment were similar in both groups. No systemic adverse events were reported; local side effects were similar in terms of intensity and frequency. Therefore, it has been concluded that onabotulinumtoxinA and the Chinese product have similar efficacy, safety and tolerability profiles, so that a dose equivalence of 1:1 may be considered for blepharospasm treatments [43]. These results need to be replicated in larger series, as the quantity and quality of data supporting the observation are limited.

In addition to BoNT/A, rimabotulinumtoxinB has also been used successfully in the treatment of blepharospasm [44], but double-blind controlled studies in this disorder are lacking. In a retrospective review of 16 patients resistant to BoNT/A and treated with BoNT/B, the mean effect equaled 7.3 weeks and was rated as fair to excellent in the majority. However, in this study, side effects were common and included pain at the site of injection, ptosis and dry mouth. Switching to an alternative BoNT serotype may benefit "secondary nonresponder" patients (those who have initial clinical benefit from BoNT injections that wanes over time) [45].

Understanding the muscular anatomy is critical to ensure optimal results. Various BoNT injection techniques have been advocated to optimize response and minimize adverse effects. The standard treatment techniques involve injection into four sites around each eyes, two in the upper lid, one medially, and one laterally near the canthus. Two additional injection sites in the lower lid, one at the lower lateral canthus and one near the lower lid midline, seem to produce a longer duration of effects than those in the eyebrows, inner orbital and outer orbital [46]. Blepharospasm may differentially affect the three concentric parts of the orbicularis oculi muscle; inadequate results are obtained if the toxin is injected in the orbital portion of a patient suffering from a predominant involvement of the pretarsal portion of the muscle [29]. In a retrospective study of 25 patients with blepharospasm, compared to preseptal placements, pretarsal BoNT/A injections produced a better response rate with a longer duration and lower incidence of ptosis, the most common side effect [31]. It was concluded that pretarsal placement is sufficient to provide optimum results, leaving the option to add preseptal or orbital injections if necessary. Furthermore, patients with a predominant pretarsal involvement may have prevalently tonic eye closure and find it difficult to voluntarily open the eyelids (so-called eyelid-opening apraxia). In such cases, EMG recordings show loss of the normal reciprocal inhibition between the levator palpebrae and the pretarsal portion of the orbicularis oculi, with co-contraction. BoNT/A is helpful in these cases if injected in the pretarsal portion and at doses lower than the ones used in the orbital part of the muscle [28]. Other muscles that may also be involved in blepharospasm include the corrugator supercilii, the frontalis and the procerus.

3.2.3 Cervical Dystonia

Cervical dystonia is the most common form of primary focal dystonia, also referred to as "spasmodic torticollis"; its incidence has been estimated in 5–9/100,000 [47] and prevalence in 20–200 per million [48]. It is a neurologic condition that causes abnormal movements and postures of the neck. The phenomenology of cervical dystonia is complex; it can variably combine tonic (slow and sustained) and phasic (fast and intermittent) movements. Overlying spasms can induce slow and rapid head jerks. Cervical dystonia arises from involuntary activation of muscles causing turning (torticollis), tilting (laterocollis), flexion (antecollis) or extension (retrocollis) of the head; sometimes these are combined with elevation or anterior shifting of the shoulder. Each of these postures is associated with specific patterns of muscle overactivity in each patient, with variability from patient to patient. Pain affects approximately 60 % of cervical dystonia patients and can be the most disabling feature. Cervical dystonia most commonly presents as a sporadic disorder of adulthood, but up to 12 % of patients may report a positive family history [49].

Commonly, cervical dystonia starts in the 40s; it is a lifelong condition; permanent remissions are rare, although temporary remissions lasting days to years may occur [50]. Although not life threatening, cervical dystonia can cause disability and impair quality of life [51]. Moreover, several disabling conditions (cervical arthritis, radiculopathy, and myelopathy) may occur concomitantly [52]. Secondary cases of cervical dystonia have also been described following cervical or brain traumatic injury, or in association with neurodegenerative diseases or cerebral palsy (CP). The assessment and treatment of secondary forms of cervical dystonia have not been subject to the same rigorous studies as primary focal cervical dystonia.

Several treatment options are available for cervical dystonia, including oral pharmacological agents, soft tissue surgery, surgical denervation and deep brain stimulation (DBS). Oral medications (anticholinergic agents, baclofen and benzodiazepines) may be of limited benefit; their use is limited by common side effects. Although the use of DBS in patients with dystonia is recent, there is growing evidence that globus pallidus DBS is an option for patients with severe symptoms [53], [54]. Among the therapeutic interventions available, BoNT is regarded as the first-choice treatment, due to its efficacy and positive cost/benefit ratio. Up to 85 % of patients get benefit from BoNT treatment, particularly as it concerns ameliorating head posture, reducing pain and improving range of motion. For these reasons, this has long been considered the treatment of choice in cervical dystonia patients [55], [56].

Since the first report of the efficacy of the original North American BoNT/A batch (Oculinum®) [57], more than 80 studies (mostly uncontrolled or consisting of small series) have evaluated BoNT in cervical dystonia. Among these, eight prospective, double-blind, randomized controlled clinical studies provided class I evidence of the efficacy of BoNT in ameliorating head posture and neck pain (8). Other studies documented the improvement of health-related quality of life and disability after BoNT. One study compared abobotulinumtoxinA injections with oral administration of trihexyphenidyl and found that BoNT/A is more efficacious with fewer adverse events [58].

All commercially available BoNT brands have proven efficacious in randomized controlled trials (RCTs) on cervical dystonia. Notwithstanding this evidence, several questions remain unresolved. The first is whether the three BoNT/A brands are equivalent and the second is what place the BoNT/B formulation has in the treatment algorithm. These products are not identical, in either formulation or dose [2] and there are suggestions of potential differences in efficacy and safety profiles among BoNT preparations.

In clinical practice, when shifting from one brand of BoNT/A to another or from BoNT/A to BoNT/B, there is no clear dosing equivalency [59]. Dosing in cervical dystonia patients varies depending on serotype and brand. Two prospective studies compared onabotulinumtoxinA to abobotulinumtoxinA. In one blinded, parallel-arm study, a fixed dose ratio of 1 onabotulinumtoxinA U to 3 abobotulinumtoxinA U showed similar efficacy, adverse effect profile and duration [60]. However, a subsequent study from the same group did not confirm this observation and reported that the dose equivalency of onabotulinumtoxinA and abobotulinumtoxinA was less than 1:3 [61]. Furthermore, a retrospective study analyzing patients switched from abobotulinumtoxinA to onabotulinumtoxinA or vice versa found a variable dosing ratio ranging from 3 to 5:1 [62]. This suggests that different brand units cannot be converted linearly. It is therefore recommended that each BoNT/A brand be administered according to the dosing suggestion of the information package and the patient's needs. In cervical dystonia, onabotulinumtoxinA doses vary between 70 and 370 U. Doses < 100 U are usually sufficient to relieve cervical pain in the majority of patients [63]. As for abobotulinumtoxinA, it has been recommended to start with a dose of 500 U that provides benefit in most patients with minimal risk of adverse events [64]. A study comparing 250, 500 and 1,000 abobotulinumtoxinA U in cervical dystonia reported that the magnitude and duration of improvement was greatest after injections of 1,000 U, at the cost of significantly more adverse events [65]. IncobotulinumtoxinA has been compared to onabotulinumtoxinA in a non-inferiority trial reporting that this BoNT/A brand is as efficacious and safe at a

1:1 dose ratio as onabotulinumtoxinA [66]. Direct comparison of onabotulinumtoxinA and rimabotulinumtoxinB was performed in two studies that established a dose ratio between 1:40 and 1:66.6 U. Both studies showed comparable efficacy. In the first study, the onabotulinumtoxinA-treated group had a modestly longer duration of benefit (approximately 2 weeks) and fewer occurrences of dysphagia and dry mouth than the rimabotulinumtoxinB group (6). In the second study, there was no difference in duration or adverse events [67].

A Chinese BoNT/A brand (Prosigne®) has been compared to onabotulinumtoxinA® in a prospective, randomized, double-blind study. Average duration of effect and incidence of adverse events were similar; social aspect, pain and quality of life improved in both groups; the authors concluded that these two BoNTs were equivalent in terms of efficacy, safety and tolerability profiles, with a dose equivalence ratio for cervical dystonia of 1:1 [43]. As for blepharospasm, more experience and higher quality trials are needed for this toxin brand. BoNT/F has also been shown to improve cervical dystonia symptoms in secondary nonresponders. However, the benefit duration is much shorter, lasting for approximately 8 weeks [68]. Increasing the dose prolongs the duration of clinical benefit at the cost of increased adverse effects [69]. It has also been observed that after repeated injections approximately 33 % of these patients developed resistance to serotype F [68].

In clinical practice, the average total dose injected in patients with cervical dystonia is 100–300 onabotulinumtoxinA U or incobotulinumtoxinA U, 400–800 abobotulinumtoxinA U, or 10,000–20,000 rimabotulinumtoxinB U. These doses can vary considerably as the recommended range has to be adjusted depending on the individual patient's features. It is also generally accepted that larger doses are associated with an increased risk of adverse events [70]. The initial treatment should be targeted to the most active muscles contributing to dystonic movements and postures. Most studies report that the average latency of clinical action is around 1 week. The benefit duration is reported to last between 8 and 16 weeks, although it may be as long as 5–6 months, especially with repeated sessions. However, on average the patients need re-treatment every 3–4 months. Duration of benefit has been observed to last longer in patients with moderate cervical dystonia [71]; efficacy on pain reduction is more marked than that on involuntary movements.

Adverse events are generally mild or moderate and transient, including pain at injection site, neck weakness, flu-like symptomatology, hoarseness, dry mouth and dysphagia [72]. Systemic events include general tiredness and muscle weakness (occurring even in the placebo arm of controlled studies) [73]. Differences in adverse-event rate among BoNT preparations may be important for selecting a treatment and setting expectations. After injection of cervical muscles, the most severe side effect and dose-limiting factor is dysphagia, caused by migration of BoNT out of the injected muscle. According to some studies, abobotulinumtoxinA is more efficacious than onabotulinumtoxinA in controlling pain and dystonia [61], but has a higher incidence of side effects (dysphagia, dysphonia, asthenia, neck weakness), probably because of a higher diffusion around the injection sites [74]. Dysphagia and dysphonia are considered the two most important side effects related to BoNT diffusion to the underlying pharyngeal and laryngeal muscles after injection in the sternocleidomastoid muscle. Particular care should be taken to avoid diffusion outside the sternocleidomastoid towards deeper structures, limiting the doses [65] and choosing the appropriate dilutions and injection sites. There is evidence that autonomic dysfunction is more common with BoNT/B compared to BoNT/A [75]. This has been confirmed by two class I studies [6], [67]. Dry mouth is commonly associated with BoNT/B injection and seems unrelated to the doses injected, presumably because BoNT/B blocks the cholinergic release in postganglionic parasympathetic fibers to the salivary glands (Table 3.3).

Patients with cervical dystonia who do not improve after BoNT treatment are called primary nonresponders; those who do not improve following a previous successful treatment are called secondary nonresponders. Primary failure occurs in approximately 15–30 % of patients and has several causes, including contractures, inadequate dosing, inaccurate muscle selection, inaccessibility of the muscles involved or patient's immunization. For example, in patients with antecollis, BoNT injections may be unsuccessful because of the involvement of prevertebral muscles that are not accessible for injection. Retrospective studies suggest that secondary failure to BoNT affects approximately 10–15 % of patients with cervical dystonia [72], [76]. The occurrence of antibodies to BoNT, revealed by the mouse neutralization assay, has been reported in one third of secondary nonresponding patients [76]. In patients who develop resistance to one serotype, treatment with another serotype may restore clinical efficacy [77]. It is advisable that the frequency of repeated treatments is reduced as much as possible to minimize the risk of immunization.

3.3 Other Focal Dystonias

3.3.1 Oromandibular and Lingual Dystonia

Oromandibular dystonia (OMD) is a focal form that mainly involves the masticatory muscles and also affects the lower facial, labial and tongue muscles. Masticatory muscles spasms can induce jaw closing or opening, lateral deviation, protrusion, retraction, or a combination of different movements. Involuntary biting of the tongue, cheek, or lips and difficulty in speaking and chewing is often socially embarrassing and cosmetically disfiguring. Lingual dystonia often occurs in association with other OMD forms, but can be isolated as well. It is rare and disabling, impacting daily activities (e.g., speaking, chewing, swallowing) and causes social disability.

OMD affects women more than men. The mean age at onset is between 50 and 60 years [78]. The picture tends to remain stable, but fluctuations are observed in individual cases. Although spontaneous improvement may occur with time, complete remissions are exceptionally rare. Dystonia in OMD is commonly worsened by action, in particular with specific motor tasks, such as eating or praying [79].

Most patients with OMD have a primary condition, while tardive dystonia represents the most common cause of secondary OMD. Trauma or procedures involving

Table 3.3 Class I studies on BoNT treatment in neuromuscular disorders (classification based on American Academy of Neurology criteria)

Disease	Study design	BoNT type	Dose (U total)	Patients (n)	Reference
Blepharospasm	Double-blind randomized, non-inferiority	OnabotulinumtoxinA vs. IncobotulinumtoxinA	40.4 U (mean)	300	[39]
Cervical dystonia	Double-blind, randomized, placebo controlled	OnabotulinumtoxinA	150–165 U	55	[245]
Cervical dystonia	Double-blind, randomized, controlled, prospective	AbobotulinumtoxinA vs. Trihexyphenidyl	554 U vs. 16.5 mg (mean)	66	[58]
Cervical dystonia	Double-blind, placebo-controlled	RimabotulinumtoxinB	2,500/5,000/10,000 U	122	[246]
Cervical dystonia	Double-blind, randomized, placebo controlled	AbobotulinumtoxinA	250/500/1,000 U	75	[65]
Cervical dystonia	Double-blind, randomized, parallel	OnabotulinumtoxinA vs. AbobotulinumtoxinA	159 U vs. 477 U (mean)	73	[60]
Cervical dystonia	Double-blind, placebo-controlled, randomized	RimabotulinumtoxinB	5,000/10,000 U	109	[247]
Cervical dystonia	Double-blind, placebo-controlled	RimabotulinumtoxinB	10,000 U	77	[77]
Cervical dystonia	Double-blind, randomized, placebo-controlled	AbobotulinumtoxinA	500 U	80	[248]
Cervical dystonia	Double-blind, randomized, parallel-group	OnabotulinumtoxinA vs. RimabotulinumtoxinB	250 U vs. 10,000 U	139	[6]
Cervical dystonia	Double-blind, randomized	OnabotulinumtoxinA vs. RimabotulinumtoxinB	150 U vs. 10,000 U	111	[67]
Writer's cramp	Double-blind, randomized, placebo-controlled	AbobotulinumtoxinA	178 U (mean)	40	[122]
Upper limb spasticity	Double-blind, randomized, placebo-controlled	OnabotulinumtoxinA	75/150/300 U	39	[249]
Upper limb spasticity	Double-blind, randomized, placebo-controlled	AbobotulinumtoxinA	1,000 U	24	[250]
Upper limb spasticity	Double-blind, randomized, placebo-controlled	AbobotulinumtoxinA	500/1,000/1,500 U	25	[251]
Upper limb spasticity	Double-blind, randomized, placebo-controlled	AbobotulinumtoxinA	1,000 U	40	[252]
Upper limb spasticity	Double-blind, randomized, placebo-controlled	AbobotulinumtoxinA	500/1,000/1,500 U	82	[253]
Upper limb spasticity	Double-blind, randomized, placebo-controlled	AbobotulinumtoxinA	1,000 U	59	[254]
Upper limb spasticity	Double-blind, placebo-controlled	OnabotulinumtoxinA	200–240 U	122	[161]
Upper limb spasticity	Double-blind, randomized, placebo-controlled	OnabotulinumtoxinA	90/180/360 U	91	[255]
Upper limb spasticity	Double-blind, randomized, placebo-controlled	RimabotulinumtoxinB	10,000 U	15	[172]
Upper limb spasticity	Double-blind, randomized, placebo-controlled	AbobotulinumtoxinA	350/500/1,000 U	50	[256]
Upper and lower limb spasticity	Double-blind, randomized, placebo-controlled	OnabotulinumtoxinA	141.25 U (upper limb), 284.75 U (lower limb), (mean)	52	[257]

Table 3.3 (continued)

Disease	Study design	BoNT type	Dose (U total)	Patients (*n*)	Reference
Lower limb spasticity (hip adductor)	Double-blind, randomized, placebo-controlled	AbobotulinumtoxinA	500/1,000/1,500 U	74	[177]
Lower limb	Double-blind, randomized, placebo-controlled	AbobotulinumtoxinA	500/1,000/1,500 U	234	[258]
Spastic equines (CP)	Double-blind, randomized, placebo-controlled	OnabotulinumtoxinA	4 U/Kg	20	[259]
Spastic equines (CP)	Double-blind, randomized, placebo-controlled	AbobotulinumtoxinA	15/25 U/Kg	40	[260]
Spastic equines (CP)	Double-blind, randomized, placebo-controlled	AbobotulinumtoxinA	10/20/30 U/Kg	126	[261]
Adductor spasticity (CP)	Double-blind, randomized, placebo-controlled	OnabotulinumtoxinA	8 U/Kg	16	[262]
Adductor spasticity (CP)	Double-blind, randomized, placebo-controlled	AbobotulinumtoxinA	30 U/Kg (max 1,500 U)	61	[263]
Lower limb spasticity (CP)	Double-blind, randomized, placebo-controlled	OnabotulinumtoxinA	12 U/Kg (max 400 U)	33	[264]
Upper limb spasticity (CP)	Double-blind, randomized, controlled	OnabotulinumtoxinA	4.6 vs. 9.2 U/Kg	39	[265]
Leg spasticity (CP)	Double-blind, randomized, placebo-controlled	AbobotulinumtoxinA	15–30 U/Kg	64	[266]

the face or oral and dental structures have been suggested to be causative [80]. Occasionally, OMD has been observed as an accompanying manifestation of neurodegenerative disorders, focal brain lesion or brainstem lesion [81]. Finally, OMD can lead to secondary complications, such as tension-type headache, increased dental wear, temporomandibular joint syndrome or temporomandibular joint dislocation. In order to prevent these complications, early diagnosis and appropriate treatment are crucial.

The presentation of OMD is highly variable and treatments need to be individualized. Pharmacological therapy is only partially effective [82]. Oral medications, including anticholinergics, tetrabenazine, baclofen or clonazepam, can be used. Tetrabenazine, in particular, is helpful in lingual protusion dystonia [83], [84]. Muscle afferent block by intramuscular injection of lidocaine and alcohol has been shown to be helpful, but further experience and evaluation are needed to determine its long-term efficacy and benefit [85]. Lastly, pallidal DBS has been performed in a few patients with positive results and may be considered as an option in some patients with intractable OMD [86], [87].

3.3.1.1 BoNT Treatment in OMD and Lingual Dystonia

BoNT has become the therapy of choice for OMD, and its use in jaw-opening, jaw-closing and jaw-deviation OMD has been documented [88], [89], although most data derive from open studies. The best responses have been reported on jaw-closing OMD [78].

In jaw-closing and jaw-deviation dystonia, BoNT is injected into both masseters and temporalis muscles. Typical doses in the masseters are 25 onabotulinumtoxinA or 100 abobotulinumtoxinA units; in the temporalis muscles typical doses are 20 and 80 units, respectively. If these injections are not sufficient to control dystonic movements, the internal pterygoid can be injected (with 15 onabotulinumtoxinA U or 60 abobotulinumtoxinA U). Scanty data are available on rimabotulinumtoxinB [90], [91]. Suggested doses are 2,500 units in each masseter muscle and 1,000 units in the pterygoids [92]. There is no experience with incobotulinumtoxinA.

The treatment of patients with jaw-opening dystonia is more challenging; in this situation, the most important muscle to treat is the external pterygoid that can be approached transorally or laterally through the mandibular incisure. Notwithstanding, the digastric and other muscles can play a role. The external pterygoid is injected with 15 onabotulinumtoxinA units or 60 abobotulinumtoxinA units and the digastric muscle with 10 and 40 units, respectively. This combination is usually effective. In some patients, injecting the platysma with 20 onabotulinumtoxinA U, 60 abobotulinumtoxinA U or 1,000 rimabotulinumtoxinB U can provide additional improvement.

In jaw-deviation dystonia (often combined with protrusion), the contralateral external pterygoid muscle is the most important muscle to treat; when jaw-protrusion dystonia is dominating, both external pterygoids are often involved. Pterygoid muscle injections have to be performed with EMG guidance, as the muscles are not easily accessible to palpation. The use of EMG is often helpful for other jaw muscles (digastric, masseter, temporalis). BoNT may also improve the symptoms of temporomandibular joint syndrome and other oral and dental problems, as well as dysarthria and chewing difficulties. Transient swallowing difficulties have been reported in less than 20 % of treatment sessions. BoNT treatment may be disappointing in severely disabled patients for whom other solutions, such as DBS of the globus pallidum internum, have to be considered.

Lingual dystonia is difficult to treat and significant adverse effects have been reported. The need to preserve functional activity limits the amount of toxin that can be used. Especially in patients with severe tongue protrusion, results are disappointing [93]. Injections of 10 onabotulinumtoxinA U or 40 abobotulinumtoxinA U into the intrinsic tongue muscles can be used in lingual dystonia. More recently, it has been suggested that lingual protrusion dystonia may be successfully treated by injecting the genioglossus muscles. However, the risk of dysphagia is high, so it is recommended to start with very low doses (5 onabotulinumtoxinA U) in each genioglossus muscle and then increase by 2.5 U up to 15 U per treatment session until the patient achieves a reasonable response. Despite this prudent approach, dysphagia may still occur [94].

There is no reported experience with rimabotulinumtoxinB or incobotulinumtoxinA in lingual dystonia.

3.3.2 *Spasmodic Dysphonia*

Spasmodic dysphonia (SD) is a laryngeal dystonia, most often focal, that sometimes may occur in association with cranial or generalized dystonia. The vocal folds are normal at rest, but during phonation they develop action-induced, task-specific contractures causing abnormal movements and muscle spasms during speaking and resulting in dysphonia [95].

Different types of SD have been identified. The adductor type, caused by spasmodic activity of the vocal muscle (thyroarytenoid), is the most common; it induces hyperadduction of the vocal folds during speaking, producing a "strain-strangled" voice that is harsh, often tremulous, with inappropriate pitch or pitch breaks, breathiness and glottal fry. The abductor form is less common; it is due to spasms of the posterior cricoarytenoid muscles, causing a prolonged, inappropriate abduction of vocal folds during voiceless consonants. This results in a breathy, effortful, hypophonic voice with abrupt termination of voicing, aphonic or whispered segments of speech. Some retain that all patients have mixed adductor/abductor involvement with predominance of either of the two. There are also patients with compensatory or pseudoabductor forms, who whisper to compensate for the tight adductor spasms they experience. In some cases, the presentation at onset may change with time; particularly, adductor may turn to abductor.

In another rare type, the adductor breathing dystonia, there are adductor spasms during respiration. The paradoxical motion creates stridulous noises during inspiration, but usually does not produce hypoxia. Other laryngeal activities are normal.

Also a "singer's laryngeal dystonia" has been identified. In this form, the vocal abnormalities occur during singing. SD typically affects patients in their mid-40s and is more common in women [96], [97]; most often SD symptoms develop gradually over several months to years.

For many years, the only treatment options for these patients were speech therapy or psychotherapy, with poor results overall. Speech therapy alone does not yield significant improvement but combined with BoNT allows treatment of the compensatory behaviors superimposed to SD [98]. Although psychotherapy may help the patients manage the associated social stress and minimize the emotion-related voice deterioration, there is no evidence that psychotherapy or psychological intervention can relieve SD.

Occasionally patients may improve with benzodiazepines (i.e., clonazepam, lorazepam) or with baclofen, and those with superimposed voice tremor may benefit from anticonvulsants (i.e., gabapentin, primidone) or beta-adrenergic antagonists (i.e., propranolol).

Until the introduction of BoNT, surgical interventions had been the only truly efficacious options, but side effects and disappointing long-term results limited its usefulness.

3.3.2.1 BoNT Treatment in SD

The first BoNT treatment was performed in 1984 on a patient with adductor SD [99]. Following this pioneering series on adductor SD, patients with abductor SD were also treated starting in 1988 [100]. In the past two decades, enough evidence has been produced to conclude that BoNT/A (or BoNT/B if there is resistance to type A) are the first-line treatment for SD [101]. Currently, BoNT is considered the treatment of choice for this disorder; most investigators report a 75–95 % improvement in voice symptoms and a significant improvement in the quality of life. Adverse events include transient breathy hypophonia, hoarseness and occasionally dysphagia with aspiration.

Most commonly, adductor SD is treated by injecting percutaneously the laryngeal adductor muscles under EMG guidance. Unilateral or bilateral protocols have been proposed for BoNT injections into the thyroarytenoid muscle. Some groups have proposed treatment with large doses (20–30 onabotulinumtoxinA U) given unilaterally to minimize adverse events [102]. When bilateral treatments were compared with unilateral ones, the latter showed a more favorable efficacy/tolerability profile [103]. The most experienced injectors, however, retain that after an initial unilateral treatment of 2.5–7.5 onabotulinumtoxinA U, application of a bilateral protocol in subsequent treatment session prevents exacerbation of laryngeal dystonia in the untreated side [104]. Similar experience has been gathered with abobotulinumtoxinA [105]. In a prospective study, 31 patients with adductor SD were treated for five consecutive times, either unilaterally or bilaterally. Low-dose unilateral injections into the thyroarytenoid muscles produced comparable results to bilateral treatment, regarding duration, voice improvement and complications; moreover, unlike bilateral injections, unilateral ones were not associated with complete voice loss [106].

The doses of BoNT used in SD can vary depending on the toxin brand and the technique used. In the earlier literature, the doses varied from 3.75 to 7.5 onabotulinumtoxinA U for bilateral injections [107], [108] to 15 U for unilateral injections [102]. Remarkably, up to 50 onabotulinumtoxinA U in each vocal cord have been used [109]. Lower doses were later recommended [110]. Truong and colleagues suggested to start with 0.5 onabotulinumtoxinA U or 1.5 abobotulinumtoxinA U when injecting bilaterally, then to adjust the dose as needed (the estimated average dose being 0.75–1 onabotulinumtoxinA U or 2–3 abobotulinumtoxinA U) [111]. The duration of improvement is dose related. In the long term, the average latency of effect was 2.4 days with a peak at 9 days and a duration of 15.1 weeks [104]. This treatment is generally well tolerated; breathiness was often reported as transient or mild. Alternating unilateral injections caused significantly less breathy voice than bilateral injections [103]. A slightly higher incidence of aspiration, dysphagia and breathiness was reported in the bilaterally injected group of patients, who required significantly lower doses of toxin to attain benefit [112].

The experience with BoNT/B in SD is limited to the treatment of adductor-type dysphonia. In one patient who failed to respond to BoNT/A, 250 rimabotulinumtoxinB U were injected in each vocal fold with beneficial effects lasting for 3.5 months [113]. RimabotulinumtoxinB was found to be safe and effective in a class IV single-site, open-label study. It has been reported that 8 out of 10 treated patients, who received 200 rimabotulinumtoxinB U on each side, had a clinical improvement lasting for 8 weeks [114]. Three patients, who failed to respond to BoNT/A and subsequently received BoNT/B (up to 1,000 rimabotulinumtoxinB U per side), were reported to show improvement for approximately 2 months [111]. A direct comparison of BoNT/A and BoNT/B was performed on 32 patients with adductor SD who had been treated with stable BoNT/A doses and were followed up for 1 year with BoNT/B [115]. The conversion rate for laryngeal injections was considered to be 52.3:1. RimabotulinumtoxinB had more rapid onset and shorter duration of action (10.8 vs. 17 weeks). The safety profile was comparable.

Abductor SD is a difficult-to-treat condition. Usually the posterior cricoarytenoid muscles, the cricothyroid muscles or both are involved, but generally only the posterior cricoarytenoid muscle is injected under EMG guidance.

Bilateral injections are dangerous, as side effects include stridor and airway obstruction. Therefore, unilateral injections with 2.5–25 onabotulinumtoxinA U are performed on the most active side, as determined by fiberoptic laryngoscopy. A common procedure is to inject 5 onabotulinumtoxinA U into the more active posterior cricoarytenoid muscle. If there is no subjective improvement in voice quality after 2 weeks, and if no airway symptoms have occurred, then the opposite muscle is injected with an additional 5 onabotulinumtoxinA U [116]. The following procedure has also been proposed: 2–4 onabotulinumtoxinA U on the most active side with 1 U in contralateral muscles, or 12 abobotulinumtoxinA U on the most active side and 3 abobotulinumtoxinA U on the opposite side [111]. A lower dose protocol with 1.25–1.75 onabotulinumtoxinA U in one muscle and 0.9 U on the opposite side has also been implemented [117]. Generally, if a high dose is required on both sides, the second side can be injected with a delay of 2 weeks to avoid compromising the airway. Simultaneous bilateral posterior cricoarytenoid muscle injections have also been considered to be safe [118]. The total BoNT dose injected in each session was between 2.50 and 7.50 onabotulinumtoxinA U, with an average total dose per session of 4.70 U. There were no life-threatening complications.

In a 12-year long experience on 154 patients, approximately 20 % of them had significant voice improvement associated to weakening or paralysis of one posterior cricoarytenoid muscle. The remaining 80 % needed an additional dose of 0.625–2.5 onabotulinumtoxinA U into the contralateral posterior cricoarytenoid. The overall improvement was around 70 %; the onset of efficacy was on an average 4.1 days and the benefit lasted for 10.5 weeks. Side effects were observed in 2 % of patients, consisting of mild exertional wheezing and 6 % mild transient dysphagia to solids [119]. The cricothyroid muscle can also be injected, under EMG guidance, by percutaneous access. In a large series of SD patients, nine received bilateral injections (2.5 onabotulinumtoxinA U on each side) in the cricothyroid muscles in addition to treatment in the posterior cricoarytenoid. These patients still had breathy breaks despite significant limitation of abduction. Five of the nine injected cases had benefit consisting in a louder voice with fewer breaks. One patient got worse after the additional injection [104]. Therefore, BoNT is probably effective for the treatment of adductor SD but there is less evidence to support its use in abductor SD.

In summary, laryngeal dystonia is a heterogeneous condition that can be improved by BoNT. Different treatment schemes and doses are required to fit the many varieties of presentations.

3.3.3   Focal Limb Dystonia

Albeit BoNT represents the treatment of choice for focal limb dystonias, functional outcome of treatments is disappointing compared to that of blepharospasm or cervical dystonia, particularly because hand movements involve the subtle tuning of many forearm and hand muscles. Still, there are no effective alternative medical or surgical treatments. Writer's cramp (WC) in particular affects the sophisticated function of writing. As for other occupational cramps, it is difficult to obtain the requested quality of voluntary movement without weakness.

3.3.3.1   Upper Limb

The upper extremity is affected more commonly than the lower limb. Focal upper limb dystonia usually begins in the hand and is task specific; with progression, task specificity is gradually lost. Typical upper limb dystonias include musician's cramps and WC, where BoNT has been reported to be effective [120], [121].

Most studies on WC are open-label reports of clinical experiences. A class I randomized, double-blind, placebo-controlled trial in 40 patients with WC treated with abobotulinumtoxinA showed BoNT/A efficacy based on subjective and objective clinical scales [122]. Temporary weakness and pain at the injection site were the only reported adverse events. This observation has been replicated in three class II double-blind trials on upper limb injections of onabotulinumtoxinA [123], [124]. Pain is the symptom most frequently improved after treatment, often independently of motor function. The desired goals of BoNT treatment vary in each patient. The most immediate goal is to correct abnormal hand posture and relieve discomfort. The primary goal of restoring normal hand function is extremely difficult to achieve, as a consequence, despite initial improvement, some patients do not continue injections. Other patients are dissatisfied with the degree of benefit because BoNT does not fully correct all the symptoms, in particular loss of speed and coordination that is especially problematic for professional musicians. Secondary resistance due to antibody formation has been described in approximately 10 % of patients treated with the original onabotulinumtoxinA batch for focal hand dystonia [120]. Type A-resistant patients have been effectively treated with BoNT/F (68).

The first step in treatment planning is to identify the muscles most severely affected, separating out dystonic from compensatory movements. After initial inspection, EMG muscle selection usually allows to refine the choice of targets [121]. Injections can be performed using EMG- or ultrasound-based targeting.

The dose of BoNT is based on muscle size. Injections are repeated about every 3 months. In WC, the muscles injected usually include finger flexors and extensors and, if needed, also wrist pronators and flexors. Dose ranges are: 10–50 onabotulinumtoxinA U or 30–120 abobotulinumtoxinA U per muscle [125]. The importance of EMG-guided targeting is supported by the observation that only 37 % of needle placements based on surface anatomy were appropriately localized in the target muscle [126].

3.3.3.2 Lower Limb Dystonia

Foot dystonia can be either idiopathic, in the context of a generalized dystonia, or symptomatic as in Parkinson's disease (PD) or in juvenile CP. Successful treatments with BoNT have been reported but no controlled trials are available [127–129]. BoNT use is still recommended since therapeutic alternatives are lacking. Higher doses may be given than in hand dystonia because motor control is less refined.

Lower limb dystonias often present with foot inversion, toe dorsiflexion and/or ankle plantar flexion. The injected muscles may include tibialis posterior, extensor hallucis longus, gastrocnemius and long toe flexors.

3.4 Hemifacial Spasm

HFS, a form of segmental myoclonus, is characterized by involuntary, intermittent and irregular clonic twitches or tonic contractions of the muscles supplied by the facial nerve on one side of the face [130]. HFS is a sporadic disorder with occasionally familial occurrence. Some patients may be genetically predisposed to develop HFS, but most cases are sporadic [131]. It occurs more commonly in women (2:1) with an overall prevalence around 10/100,000, but in some populations, such as the Asians, the prevalence is much higher [132].

Most cases of HFS are attributed to an aberrant or ectopic artery (anterior inferior cerebellar, posterior cerebellar or vertebral) compressing the facial nerve at the root exit zone, resulting in an axono-axonal "ephaptic" transmission and a hyperexcitable facial motor nucleus. However, up to 25 % of unaffected individuals have vascular loops compressing the facial nerve, suggesting that this phenomenon alone may be insufficient to cause HFS [133]. Reports which have been associated to HFS include meningioma, schwannoma, neurinoma of the acoustic nerve, parotid gland tumor and pilocytic astrocytoma of the fourth ventricle. These space-occupying lesions should be excluded, in particular in patients with atypical features such as facial weakness or decreased corneal reflex, or any other evidence of cranial nerve dysfunction. Sometimes, peripheral facial nerve injury or prior Bell's palsy can also precede HFS; in those cases, hyperkinesias often coexist with a mild ipsilateral facial weakness [134]. Patients without history of Bell's palsy still may have abnormal EMG findings suggesting an old facial nerve damage and subsequent pathological regeneration [130].

Most patients present with unilateral contractions, but bilateral cases of HFS have been reported [135], [136]. Usually, the disorder starts in the orbicularis oculi muscle and gradually spreads to other muscles, such as the frontalis, procerus, zygomaticus, risorius, levator labii superioris, depressor labii inferioris, depressor anguli oris and sometimes also platysma. Although HFS is not a life-threatening condition, it may have a severe impact on the patient's aesthetics and causes social disability; moreover, it sometimes interferes with sleep. Rarely, patients with HFS may spontaneously remit; most require lifelong treatment.

Treatment options are aimed to reduce or stop muscular twitches and include medications, BoNT injections, neurosurgery and doxorubicin chemomyectomy.

Several symptomatic drugs have been tried. Anticonvulsant medications (such as carbamazepine, clonazepam, phenytoin, gabapentin or valproate) have been reported to improve HFS and to provide mild symptom relief. Among these, carbamazepine is the most frequently used; it has been reported to alleviate HFS in approximately 50 % of patients [137]. However, medications are often ineffective in the long-term management and side effects may be relevant [138].

A potentially curative approach is provided by microvascular decompression aimed at separating the aberrant artery from the facial nerve. This technique has a high success rate (from 88 to 97 %), and in the majority of cases resolution of HFS is durable, supporting the indication of surgery in younger patients [139], [140]. On the other hand, symptoms recur in as many as 25 % of patients within 2 years after surgery; moreover, complications occur in more than 20 % of the patients, sometimes serious, including permanent deafness, facial palsy, excessive bleeding and even death [141], [142]. Chemical rhizotomy of the facial nerve with doxorubicin is a potential alternative which has provided promising results. The most frequently reported adverse event is skin inflammation [143].

The introduction of BoNT as a therapeutic agent has represented a major milestone in the effective clinical management of HFS. AbobotulinumtoxinA was first used in HFS with appreciable results in 6 patients [144]. Based on the experience collected over the past two decades, BoNT has emerged as the first-choice option for the symptomatic management of HFS [26]. Although experience with BoNT mainly originates from open-label trials, there is no doubt on its efficacy and safety in the long term [34], [145]–[150]. Two RCTs [151], [152] and more than 30 open-label studies, encompassing overall more than 2,200 patients, are available on the use of BoNT/A in HFS. However, as pointed out from a recent Cochrane meta-analysis, the peculiarities of the different BoNT formulations, such as long-term efficacy, safety and immunogenicity, still need to be investigated [153].

A single-blind, randomized, parallel-design study comparing onabotulinumtoxinA and abobotulinumtoxinA failed to show differences in efficacy and tolerability using a 1:4 conversion rate in HFS [35]. It has been anecdotally reported that shifting from onabotulinumtoxinA to abobotulinumtoxinA may relieve HFS in secondary nonresponders [38], but this observation has not been confirmed by controlled trials. We recently performed a retrospective evaluation of outcome predictors, efficacy and safety of onabotulinumtoxinA and abobotulinumtoxinA in more than 100 HFS patients followed for a 10-year period and observed the following differences [154]. The mean duration of clinical improvement was higher after the injection of abobotulinumtoxinA than onabotulinumtoxinA by approximately 20 days (105.9 ± 54.2 vs. 85.4 ± 41.6 days, respectively, $p < 0.01$). Over time, the duration of clinical benefit slightly increased with onabotulinumtoxinA, but remained constant with abobotulinumtoxinA; ptosis and lagophthalmos were more common with abobotulinumtoxinA treatments ($p < 0.005$). This supported the view that, although both brands bear the same indications for HFS, they should be considered as two different products.

There is also experience with BoNT/B in HFS. The clinical effects lasted for about 8.5 weeks in two patients treated with rimabotulinumtoxinB over six consecutive sessions [91]. Doses ranging from 200 to 800 BoNT/B U are considered appropriate in HFS [155], but further studies are needed as the experience with BoNT/B in HFS is quite limited.

In a recent study, 17 patients with HFS, who were previously treated with onabotulinumtoxinA, were blindly converted to incobotulinumtoxinA with a 1:1 conversion rate and treated continuously for 3 years without evidence of any differences in outcome or safety profile [156]. Small studies have also assessed the efficacy of the Chinese BoNT/A brand (Prosigne®) [42], [43] and BoNT/C [157].

The injection technique plays a critical role with regard to clinical response in HFS patients. Injections are placed subcutaneously and the orbicularis oculi is easily reached by the local diffusion of BoNT; EMG guidance is not needed. The injections are placed in the orbicular or pretarsal portion of the eyelids, divided into three to four sites. Most investigators favor targeting the pretarsal portion, considering that the outcome is better (higher response rate, longer duration of response and a lower frequency of side effects), compared with preseptal injections [31]. Treatment of the periocular region leads to improvement also in the lower facial muscles (probably due to local diffusion of the toxin) [148]. At first, the extra-orbicular regions are not injected, but later, if required, these and other sites (e.g., the medial eyebrow, procerus, corrugator, frontalis muscle or the paranasal portion of the zygomaticus major muscle) can be treated. Still, if lower facial muscles are particularly active, or if there is residual contraction of the mouth following treatment in the orbicularis oculi, treating other muscles (e.g., the orbicularis oris, levator angularis, risorius, buccinator, depressor anguli oris or the platysma) should be considered.

In most studies, the average total dose used varies from 12.5 to 60 onabo-tulinumtoxinA U. from 10 to 160 abobotulinumtoxinA U or from 200 to 800 rimabotulinumtoxinB U. A prudent approach is necessary in cases of post-paralytic HFS. It is considered that a minimum threshold BoNT dose is necessary to obtain benefit, particularly for the first treatment session. In subsequent sessions, BoNT doses need to be increased or reduced based on the patient's response.

The patients usually improve soon after the first treatment session; primary or secondary failures are very uncommon. The average latency of clinical benefit varies from 2 to 6 days, and the overall response to treatment is satisfactory with a successful outcome in 66–100 % of patients. Patients with HFS have the lowest incidence of resistance to treatment, probably due to the low dosages used. The mean duration of benefit varies between 10 and 28 weeks. In most cases, the duration of efficacy increases with repeated treatments, more rarely it decreases or remains unchanged. It has also been observed that the duration of benefit is shorter in severe cases than in those of moderate severity. Prolonged remissions may spontaneously occur in a minority of patients, after a variable number of years of treatment [158].

The treatment is generally well tolerated; side effects occur in approximately 30 % of the patients and consist mainly of erythema, ecchymosis of the injected region, dry eyes, mouth droop, ptosis, facial weakness or edema. These are usually transient and resolve within 1–4 weeks. In several series, facial weakness is the most commonly reported side effect, occurring in 75–95 % of cases, mostly after injections in the mid-facial or lower facial muscles [158], [159]. Ptosis may occur following injections into the orbicularis oculi, particularly if the injection sites are too medial, abutting the levator palpebrae superioris muscle. Mild symptoms of exposure keratitis (lacrimation and irritation of conjunctiva) occur in less than 4 % of treatments, presumably due to a decreased blink rate and incomplete eye closure.

3.5   Spasticity

Spasticity is defined as a velocity-dependent increase in tonic stretch reflexes (muscle tone) that arises from abnormal processing of sensory afferent inputs to the spinal cord. Spasticity is a positive sign of the upper motor neuron (UMN) syndrome, that is a chronic motor disorder caused by UMN lesions. It is a consequence of an insult to the brain or spinal cord, which can lead to life-threatening, disabling and costly consequences. It is a central disorder of muscle tone characterized by increased resistance of an initially passive limb to externally imposed joint motion. Increased tone is a reflection of the loss of descending inhibitory (reticulospinal) influences resulting in increased excitability of dynamic fusimotor (gamma) and alpha neurons.

Besides increased tone, spasticity presents typically with increased muscle stretch reflexes, muscle spasms and clonus, weakness (spastic paralysis), and impairment of voluntary movements. Spasticity leads to exaggerated reflexes, posturing (so-called spastic dystonia), and flexor or extensor spasms, often painful. Late consequences of spasticity include contracture, fibrosis, tendon shortening and muscle atrophy.

Spasticity is frequently classified by its distribution into generalized, multifocal and focal ones. Spasticity may occur in diffuse or focal pathological disorders of the brain and spinal cord, such as stroke, multiple sclerosis (MS), traumatic brain injury, spinal cord injury and CP.

The goal of spasticity treatment is to reduce motor overactivity in order to improve movement without worsening weakness (paresis). In addition, reducing antagonist muscle overactivity may uncover functional residual power. The therapeutic approach to spasticity requires a comprehensive and multidisciplinary judgment of functional goals. The time elapsed between the acute event leading to spasticity and comprehensive patient management influences the long-term clinical picture. Successful spasticity management requires a multi-professional task force. All medical and surgical treatments need to be combined with physical interventions; therefore, BoNT injections cannot be regarded as a solo approach [160].

BoNT provides an important tool within a rich armamentarium (including physical therapy, orthosis, medication, etc.) to assemble individualized treatment plans for patients with UMN syndrome. BoNT is indicated not only to prevent and limit the functional impairment caused by spasticity, but also to provide functional improvement [161]. Safety and efficacy data lead BoNT injections to be considered as the pharmacological treatment of choice in focal spasticity, to improve limb position and functional ability and reduce pain [162].

Most studies of BoNT in limb spasticity used electrophysiological or ultrasound techniques to optimize muscle localization for injection, similarly to focal limb dystonia. A common approach is also to perform electrical stimulation or EMG targeting. EMG is not necessary for large, superficial, easily visible muscles, but is advisable for smaller and deep muscles and particularly applies to forearm and lower leg muscles, hip flexors (psoas major) and small inaccessible muscles around the jaw. The use of ultrasonography for locating both superficial and deep muscles is growing, as it is safe, noninvasive and less distressing than EMG.

The amount of toxin injected into individual muscles depends on the toxin brand, the muscle size, the number of nerve terminals located in the muscle, the number of muscles involved, the patient's age, the severity of spastic contraction and the patient's weight [163]. BoNT doses used in spasticity are higher than those used to treat other movement disorders and the upper dose limits have raised caution, particularly in children. In children, doses of 6 onabotulinumtoxinA or incobotulinumtoxinA U/kg (body weight) should not be exceeded in each muscle, with a maximum total dose of 29 U/kg [164]. A safe upper limit for abobotulinumtoxinA is 30 U/kg, with a maximum total dose of 1,000 U per child [165]. The absolute maximum abobotulinumtoxinA doses for adult have not been established, but they should probably not exceed 2,000 units in each session. A safe starting dose for children treated with rimabotulinumtoxinB is considered 400 U/kg weight that can be gradually increased to a maximum total dose of 10,000 U [166]. However, up to 17,500 rimabotulinumtoxinB U have been reported [167].

The duration of action may be appreciated 6 weeks after injection and for up to 9–12 weeks [168]. The exact timing between treatment sessions is variable; some information can be derived from experience in hyperkinetic movement disorders, but the clinical effects in spasticity may last longer than in dystonia, resulting in an average interval between treatments of approximately 3–5 months.

BoNT/A dilution affects treatment efficacy, although there are currently no recommendations on how to dilute different BoNT/A brands in spasticity. A controlled study has shown that treatment efficacy of onabotulinumtoxinA on biceps brachii spasticity may vary with changing dilutions [169]: A higher dilution results in larger injection volumes and greater neuromuscular block, probably because of more easy spread to neuromuscular endplates remote from the injection site.

BoNT has been used to treat spasticity associated with juvenile CP, cerebral stroke, brain trauma, amyotrophic lateral sclerosis or MS, but virtually every condition could be treated, as BoNT decontracts the muscles independently from the cause. The approved indications generally are, with differences from one country to the other, upper or lower limb spasticity (regardless of the etiology) and lower limb spasticity due to CP. Practical management may be as simple as injecting few muscles involved in adult-onset focal spasticity or involve a complex stepped approach as for some cases of childhood-onset spasticity requiring gradual tuning. In all cases, physical treatments are appropriately combined with BoNT injections.

Recent systematic reviews have concluded that BoNT is effective in reducing upper limb spasticity in adults and reduces muscle overactivity in a dose-dependent manner [170]. BoNT efficacy is better established for spasticity in the upper, rather than lower, limb. A limit of current evidence is that, particularly for the case of poststroke spasticity, functional improvement in patients treated with BoNT has not been investigated in detail [168]. It is believed that some disabilities related to upper limb passive and active function can improve, while the functional outcome after treatment of lower limb spasticity is poorly known. Spastic extension of the lower limb, in particular, supports standing and walking, functions that may be affected by BoNT treatments.

As with movement disorders, BoNT/A is well tolerated and safe in patients with spasticity: adverse events are limited and rare. Common side effects observed, in adults as well in children, include muscle soreness, pain at injection site, skin rash, fatigue, excessive weakness, influenza-like symptoms, infection and allergic reaction, but are generally reported to be mild and reversible. One study revealed that the most frequent problem in patients with poststroke spasticity is nausea, affecting only 2.2 % of cases [171].

There are limited data on the efficacy of BoNT/B in spasticity. One placebo-controlled trial failed to show efficacy [172] and revealed that dry mouth was a common side effect. This study also confirmed observations from treatment of dystonia patients that dose-dependent autonomic side effects are common following treatment with BoNT/B.

Future research will highlight some of the unanswered issues in spasticity treatment, such as long-term efficacy and safety and cost-effectiveness. There is also need for good quality studies on lower limb spasticity. Finally, the timing of BoNT treatment needs to be associated to treatment outcome and stratified by adjunct management strategies, such as physical and orthopedic interventions.

In adults, spasticity results from diverse etiologies, including stroke, trauma, MS, neoplasm involving the central nervous system (CNS) and amyotrophic lateral sclerosis. For the latter indication, there is insufficient documentation to assess the efficacy and safety of BoNT treatment.

3.5.1 Poststroke Spasticity

In adults, stroke is the most common cause of UMN syndrome. These patients often present postural patterns characterized by shoulder adduction, elbow and wrist flexion in the upper limb, and hip adduction, knee extension and ankle plantar flexion in the lower limb.

Although most hemiparetic patients are able to reach different ambulatory levels with rehabilitation efforts, upper and lower limb spasticity can impede activities of daily living, personal hygiene, ambulation, and in some cases, functional improvement. Paresis and increased muscle tone can also cause joint stiffness leading to contractures.

Observational and controlled studies have shown that BoNT/A improves function and symptoms in adult patients with upper or lower limb spasticity following stroke. The efficacy of BoNT/B on poststroke upper limb spasticity has been observed in open-label series, but not confirmed by controlled trials. BoNT is employed as focal antispastic agents usually as part of complex rehabilitation regimes.

There is evidence that BoNT/A is superior to placebo in reducing upper and lower limb spasticity after stroke [173]. Notwithstanding the reduction in muscle tone, there was no overall effect on functional parameters of disability. The different studies are difficult to compare, as they use different outcome measures to assess functional parameters. Reduction of hypertonia is maintained for a longer time in distal than in proximal muscles, probably due to insufficient doses injected into the larger proximal muscles [168].

A recent Japanese study on a new BoNT brand assessed the treatment of lower limb poststroke spasticity in a large, placebo-controlled clinical trial. One hundred twenty patients were randomized to a single treatment with BoNT/A or placebo, injected into lateral and medial head of the gastrocnemius, soleus and tibialis posterior muscles. This is the first large-scale large-scale trial to indicate that BoNT/A significantly reduced poststroke lower limb spasticity for 12 weeks [174].

3.5.2 Spasticity in MS

MS is the most common disabling chronic central nervous system disease among young adults and it is often complicated by spasticity. MS is a common cause of diffuse or regional muscle overactivity. In MS, it is particularly difficult to differentiate what part of functional disability is due to spasticity, and the administration of symptomatic short-lasting treatments like BoNT may contribute to define this aspect [175]. In this condition, BoNT has been used to treat thigh adductor spasticity, pes equines, striatal toe or shoulder adduction. BoNT treatment can also help patients who are bedridden or wheel-chaired and may prevent the occurrence of decubital ulcers and pain. This view has been confirmed by observational and controlled studies. Using a randomized crossover design, 400 onabotulinumtoxinA U were injected into the thigh adductor muscles; after 6 weeks, reduction of spasticity and improvement of hygiene scores have been observed without adverse events [176]. More recently, a placebo-controlled study with three different abobotulinumtoxinA doses (500, 1,000 and 1,500 U) has been performed in MS patients with hip adductor spasticity [177]. A risk–benefit assessment suggested that the optimal starting dose for treating hip adductor spasticity in MS is 500–1,000 abobotulinumtoxinA U, divided between the two legs, with subsequent dose titration as required.

Two studies evaluated the effect of BoNT/A on painful tonic spasm in MS patients. These trials showed that BoNT/A is effective in relieving pain (both the intensity and the number of painful spasms) [178]. Just as for the other indications, also in MS, physical therapy is recommended in association with BoNT treatment to improve the outcome [179]. In general, given the small numbers of MS patients studied, there is a need for further long-term studies on large cohorts. Furthermore, MS-related fatigue could be aggravated by BoNT, especially considering that large doses are needed for spasticity.

3.6 Cerebral Palsy

CP is a disorder, presenting early in life, due to prenatal, perinatal and postnatal brain injury that combines increased or decreased muscle tone, spasticity, muscle weakness, involuntary movements and loss of control of muscle coordination in various degrees. Muscle hypertonia in children combines with body growth leading to fixed contractures, torsional deformities of long bones and joint instability, which further impair the child's motor performance. Involvement of the lower limbs is responsible for early gait and balance impairment. The most dynamic developments can be observed during the first 6 years of life and all therapeutic interventions on spasticity and motor impairment must take into account the dramatic motor development taking place. Optimal therapeutic results are provided by early intervention that tap into the developmental potential of the child. Clinical manifestations may vary depending on the cause of brain injury, with spasticity being the commonest symptom.

The decision to use antispasticity medications in a child requires a careful assessment of the patient's impairment in all domains, including the occurrence of associated weakness or movement disorders, to choose the appropriate interventions. Reasons to treat spasticity include reduction of pain and muscle spasms, facilitate brace use, improve posture, minimize contractures and deformity, facilitate mobility and dexterity and improve patient ease of care as well as hygiene/self-care [180].

Pharmacologic treatment with myorelaxants and non-pharmacologic interventions such as physiotherapy and occupational therapy provide a basis to which BoNT is added. The first clinical trial with BoNT for spasticity in children with CP was reported almost 20 years ago [181]. Since then, growing evidence indicated that BoNT can decrease muscle tone and improve range in joints served by injected muscles. BoNT/A has later gained acceptance as an adjunct therapy for spasticity for children with CP. For the past 10 years, clinical experience from numerous case reports, retrospective and prospective open-label cohort studies and RCTs have described the potency of BoNT/A to treat upper and lower limb spasticity in children with CP. Additionally, several independent systematic reviews, meta-analyses and consensus statements from various groups have confirmed these observations [182].

BoNT/A combined with surgical and nonpharmacological interventions is currently the best treatment approach for children with CP. The goals of BoNT therapy go beyond a decrease in muscle tone to influence pain relief, prevention of contractures, psychological integration and global functional improvement. Scanty data are available on BoNT/B; they mostly derive from small open-label pilot studies including patients who were secondary nonresponders to BoNT/A. There is concern that, particularly in children, large BoNT doses may lead to a botulism-like symptomatology. In 2009, the US FDA ordered that the manufacturers of BoNT products add a boxed warning to the prescribing information for each product about the potential for serious side effects at sites distant from injection. The FDA also ordered the manufacturers to develop a Risk Evaluation and Mitigation Strategy, and to submit safety data on injections in children for treatment treatment of spasticity [183]. Pediatric cases involved treatment for spasticity and were described as botulism, or involved symptoms including difficulty breathing, difficulty swallowing, muscular weakness, drooping eyelids, constipation, aspiration pneumonia, speech disorder, facial drooping, double vision or respiratory depression. Serious case reports described hospitalizations involving ventilatory support and reports of death.

3.7 Tremor

Tremor is defined as a rhythmical, involuntary oscillatory movement of a body part produced by alternating or synchronous contractions of antagonistic muscles. It is the most common movement disorder and is etiologically and physiologically heterogeneous [184]. Essential tremor (ET) is the most common type of tremor and also the most commonly observed movement disorder. Propranolol and primidone usually ameliorate mild or moderate ET, but pharmacotherapy is usually not sufficient to control tremors of high amplitude that impair daily living activities.

In some of these patients, local injections of BoNT might be proposed before considering more aggressive intervention such as thalamic DBS. There are no class I studies investigating BoNT efficacy on tremor, but it is well established that this treatment is not as successful as in dystonia or spasticity [185]–[188].

BoNT/A has been tested on various tremor disorders in small open-label and controlled studies and has been proposed as a treatment for essential hand tremor [185], [186], [188]–[191]. A difficulty with the interpretation of results on tremor is that in most trials BoNT was injected according to a predetermined, rigid protocol without individualization to each patient's need, pattern and severity of tremor phenomenology. A class II class II placebo-controlled study with onabotulinumtoxinA reported improvement in tremor severity without amelioration of function and finger weakness as a side effect [191]. Another class II multicenter, randomized, placebo-controlled trial showed significant improvement of postural, but not kinetic, hand tremor in patients with ET who received 50 or 100 onabotulinumtoxinA U into the wrist flexors and extensors (188). This study provided an explanation for limited functional improvement, as it showed that kinetic rather than postural tremor is related to disability. Scant data are available for BoNT/B [91].

Primary writing tremor, a task-specific hand tremor related to focal dystonia, improved in four out of five patients treated with low onabotulinumtoxinA doses (10–12.5 U) for at least 1 year. The treatment schedule was flexible and involved the flexor carpi ulnaris, the extensor carpi ulnaris or radialis, the extensor digitorum communis and the abductor pollicis longus [192].

Data on head and voice tremor are still inconsistent. Although a number of studies reported efficacy [193]–[195], a class II study on ten patients with head tremor denied benefit [196]. In essential voice tremor, BoNT has been injected into the thyroarytenoid muscles to reduce tremor amplitude and laryngeal resistance. A class IV study suggested subjective improvement on vocal strain when speaking (195). Another class IV open-label study showed a beneficial effect of onabotulinumtoxinA in 13 patients with isolated vocal tremor and no evidence of SD [197].

Jaw tremor in PD sometimes is not controlled by antiparkinsonian medication and can improve with BoNT injections. The experience is limited to few patients, who have been treated with a mean dose of 50 abobotulinumtoxinA U (range: 30–100 U) in both masseter muscles [198]. Another case report of intermittent rapid focal jaw tremor mentioned a successful BoNT/A treatment into the masseters [199].

Palatal tremor with associated ear click may also be treated with BoNT into the tensor veli palatini muscle [200]. In these cases, BoNT should not be reserved for refractory cases, but it should be considered a safe and effective first-line therapy [201]. Tensor veli palatini, levator veli palatini or both have been injected with doses ranging between 5 and 20 onabotulinumtoxinA U or 5 and 60 abobotulinumtoxinA U. The treatment is generally safe; velopharyngeal insufficiency or nasal speech has been rarely recorded.

3.8   Tics

Tics are relatively brief, intermittent movements (motor tics), or sounds (vocal or phonic tics), usually preceded by a premonitory sensation; the association of motor and vocal tics is the clinical hallmark of Tourette's syndrome. Antidopaminergic drugs (neuroleptics) are often used to treat troublesome multifocal tics, but the risk of side effects such as tardive dyskinesias (TDs), hepatotoxicity, prolonged QT intervals, sedation and depression is quite high. Patients with focal tics affecting the eyes, the head or the larynx may be treated with BoNT/A in the affected muscles.

The first anecdotal observations were performed in patients with Tourette's syndrome and dystonic tics affecting the eyelids and neck [202]. OnabotulinumtoxinA treatment reduced the frequency and intensity of tics and ameliorated the associated premonitory sensory urge; this benefit lasted for several weeks.

Single case reports [203], [204] and case series [205]–[207] have later confirmed the improvement. One class II, double-blind, crossover study has shown that BoNT/A reduced the frequency of simple motor tics and associated premonitory urge [208]. Despite these objective improvements, the patients did not report a comparable subjective benefit from treatment, indicating the need for further evaluation of disability outcomes in tic disorders. However, a recent open-label study on 30 patients treated with onabotulinumtoxinA (2.5 U in both vocal cords) for vocal tics reported that BoNT/A also ameliorates quality of life [209]. The only relevant side effect was hypophonia. The long-term outcome of BoNT in tic disorders is still unreported.

On the other hand, BoNT has been also used to control life-threatening tics, such as dystonic cervical tics that could cause compressive myelopathy or radiculopathy [210], [211].

3.9   Other Movement Disorders

3.9.1   Tardive Dyskinesias

Drug-induced movement disorders are potentially persistent and disable abnormal involuntary movement disorders caused by exposure to dopamine receptor-blocking agents. The term "tardive" indicates iatrogenic origin related to antidopaminergic agents. The clinical features are quite variable, but most often these movement disorders present with stereotypic orolingual and facial dyskinesias that are very characteristic. There is some terminological uncertainty, as some authors use the term TDs to indicate any drug-induced movement disorder, while others mean uniquely the facial stereotyped hyperkinetic disorder and use the expression "tardive syndrome" as an umbrella term to encompass all drug-induced movement disorders. Tardive syndromes can present features of dystonia, tics, tremor, parkinsonism, akathisia, virtually the entire spectrum of movement disorders. In a minority of patients, TDs remit following withdrawal of the causative neuroleptic drug, but the hyperkinetic disorder commonly persists. Anticholinergics are prescribed in association with neuroleptics to reduce the incidence of TDs, but they are ineffective in or may aggravate TDs once these are manifest. Tetrabenazine, a dopamine-depleting drug, has been reported to improve TDs, although remission or satisfactory control of symptoms is not achieved in all cases.

Focal tardive dystonia responds to BoNT treatment as well as primary dystonia [212]. In particular, tardive blepharospasm and cervical dystonia require the same BoNT doses used to treat the primary conditions [213]–[215].

Patients with bruxism grind, gnash or clench their teeth during sleep or emotional conditions. This condition is associated with masseter (and sometimes temporalis) muscle contracture that occurs also during sleep. When severe or untreated, it can be associated with headache, dysarthria, temporomandibular joint destruction and dental wear. Bruxism may be idiopathic or symptomatic to different neurological conditions, such as parkinsonism, Huntington's disease, tardive syndromes, CP, etc. The use of night guards and other dental appliances and procedures may be helpful, but no strategies are curative. BoNT/A has been reported to be effective with satisfactory clinical control regardless of the etiology [216]. However, there are no controlled studies on bruxism. The masseter muscles (and the temporalis, when involved) have been treated bilaterally, with wide-ranging doses, from 25 to 100 onabotulinumtoxinA U.

Some cases of myoclonus have been treated with BoNT. Tinnitus associated with palatal myoclonus has proven responsive to BoNT/A (4–10 onabotulinumtoxinA U or 30–60 abobotulinumtoxinA U) injected into the tensor veli palatini muscle (or alternatively into the levator veli palatini).

Also anecdotal reports indicate that akathisia can also be treated with BoNT injections {Shulman, 1996 10703/id}.

3.9.2 Comprehensive Approach to Motor Symptoms of Parkinsonian Patients

While most of the motor symptoms, particularly the cardinal features of PD, such as tremor, bradykinesia, rigidity and gait difficulty, improve with dopaminergic drugs and other therapeutic options, including DBS, many troublesome symptoms do not respond to conventional treatments [217].

Motor symptoms amenable of treatment with BoNT include dystonia, contractures, tremor, painful rigidity and freezing of gait; non-motor symptoms include sialorrhea, seborrhea, hyperhidrosis, constipation, achalasia and overactive bladder [218].

Different forms of dystonia may complicate "on" as well as "off" periods in up to 60 % of PD patients, most often those with early onset [219], [220]. Off-period dystonia involves more frequently limbs and neck or facial muscles (mainly periocular) and can be painful, particularly in the foot [221].

Blepharospasm, apraxia of eyelid opening and oromandibular and cervical dystonia (observed not only in PD but also in other parkinsonisms, such as progressive supranuclear palsy) can be managed the same way as the corresponding forms of primary dystonia. Low starting doses of BoNT can be gradually increased until clinical benefit is achieved. BoNT may be also used to relieve pain associated to non-dystonic contractures of neck or other body regions.

Although BoNT is considered the first-line therapy in primary cervical dystonia, no class I studies proved BoNT effectiveness in cervical dystonia associated with PD. While patients with PD often have abnormal neck postures, there is some controversy whether this abnormality is due to cervical dystonia, rigidity, a combination of the two or some other mechanisms [219], [222].

Antecollis is the most common abnormal neck posture associated with parkinsonism, particularly PD and multiple system atrophy (MSA). Antecollis is difficult to treat with BoNT; moreover, the bilateral injection of sternocleidomastoid and scalenus muscles is often associated with dysphagia. The adverse effects can be avoided by a prudent approach, but treatment failures are common. The contraction of the submental muscle complex may contribute to antecollis and in some cases, an injection in this region, with or without concomitant treatment of the sternocleidomastoid and scalenus muscles, may improve the abnormal neck flexion. This approach, however, must be undertaken with great caution as dysphagia and aspiration pneumonia may complicate the treatment. By contrast, retrocollis associated to progressive supranuclear palsy can be safely and easily treated by injecting the posterior neck muscles [223].

Axial dystonia may manifest as cervical dystonia or an abnormal posture of the trunk causing scoliosis, kyphosis, camptocormia, Pisa syndrome or any combination of these. These axial features are a common cause of physical and social problems in patients with PD. BoNT has been used to treat axial postural abnormalities, including scoliosis, with uncertain results [224]–[226].

Camptocormia refers to a severe dynamic abnormal posture of the trunk with marked flexion of the thoracolumbar spine when standing and walking, almost resolved when lying in a supine position. It is associated with parkinsonian disorders such as PD or MSA [227]–[229]. The abnormal trunk flexion is often associated with EMG evidence of active contraction in the rectus abdominis. Despite the severe trunk flexion, patients with dystonic camptocormia can straighten their trunk when lying down or when raising their hands against a wall. The choice of which muscles to inject is crucial. Improvement was observed in 9 of 11 camptocormia patients who received BoNT treatment into the rectus abdominis muscle (300–600 onabotulinumtoxinA U) [228]. By contrast, ultrasound-guided injection into the iliopsoas muscles (500–1,500 abobotulinumtoxinA U on each side) was not effective in four patients with camptocormia [230].

The most common presentation of dystonia in PD is foot dystonia. Abnormal foot and hand postures may be seen in up to 10 % of untreated patients with advanced PD [231]. BoNT may be effective in correcting abnormal postures that did not yet progress to fixed contractures [232] and alleviating pain associated to peak-dose dyskinesia and end-of-dose dystonia [220]. EMG guidance may be required in order to inject deep muscles, particularly in the legs. Patients with PD and other forms of parkinsonism (such as progressive supranuclear palsy, corticobasal degeneration, etc.) occasionally develop secondary fixed dystonia of the hand which may be relieved by local BoNT injections (particularly helpful to ease pain and improve hygiene) [233].

Freezing of gait is a disabling symptom, characterized by a sudden inability to initiate gait or continue walking, particularly when facing a narrow passage, turning around or under stressful situations [234], [235]. In most cases, freezing of gait is poorly responsive to dopaminergic medication. The possibility that freezing of gait is partly due to involuntary contractions in distal muscles of legs and feet has prompted clinical trials of BoNT in PD and other parkinsonian disorders. Although initial reports were encouraging [236], [237], further observations were not confirmatory [238], [239].

Hand tremor is one of the most recognizable features of PD; it frequently interferes with the ability to hold objects such as newspaper or a cup and can often be troublesome for patients. Levodopa and other anti-PD treatments are usually effective in improving this cardinal feature of PD, but other treatment such as DBS must be considered as well. Some studies have demonstrated that BoNT may be of benefit in PD-related tremor [185], [186].

3.10 Conclusion and Outlook

BoNTs act as focal muscle relaxants and have several indications in clinical practice, particularly for the symptomatic improvement of hyperkinetic disorders. Solid evidence has been collected for different forms of dystonia and of spasticity. However, some indications still need to be supported by controlled trials. Long-term observations have proven that BoNT/A brands are safe when used by experienced doctors; caution is required when high per kilo doses are injected, particularly in children. There is much less experience with BoNT/B than with BoNT/A brands and this gap needs to be bridged.

BoNTs are useful as solo treatments (e.g., for some focal dystonia forms) or in combination with physical treatments or other procedures. Consensus algorithms need to be developed for different indications and different combination strategies, in order to facilitate homogeneity of BoNT administration among different centers and distinct specialties.

References

1. Scott AB (1980) Botulinum toxin injection into extraocular muscles as an alternative to strabismus surgery. Ophthalmology 87:1044–1049
2. Albanese A (2011) Terminology for preparations of botulinum neurotoxins: what a difference a name makes. JAMA 305:89–90

3. Albanese A, Lalli S (2009) Is this dystonia? Mov Disord 24:1725–1731
4. Frueh BR, Felt TH, Wojno TH, Musch DC (1984) Treatment of blepharospasm with botulinum toxin: a preliminary report. Arch Ophthalmol 102:1464–1468
5. Albanese A, Asmus F, Bhatia KP, Elia AE, Elibol B, Filippini G et al (2011) EFNS guidelines on diagnosis and treatment of primary dystonias. Eur J Neurol 18:5–18
6. Comella CL, Jankovic J, Shannon KM, Tsui J, Swenson M, Leurgans S et al (2005) Comparison of botulinum toxin serotypes A and B for the treatment of cervical dystonia. Neurology 65:1423–1429
7. Dressler D, Benecke R (2003) Autonomic side effects of botulinum toxin type B treatment of cervical dystonia and hyperhidrosis. Eur Neurol 49:34–38
8. Simpson DM, Blitzer A, Brashear A, Comella C, Dubinsky R, Hallett M et al (2008) Assessment: botulinum neurotoxin for the treatment of movement disorders (an evidence-based review): report of the Therapeutics and Technology Assessment Subcommittee of the American Academy of Neurology. Neurology 70:1699–1706
9. Bentivoglio AR, Fasano A, Ialongo T, Soleti F, Lo Fermo S, Albanese A (2009) Fifteen-year experience in treating blepharospasm with Botox or Dysport: same toxin, two drugs. Neurotox Res 15:224–231
10. Mohammadi B, Buhr N, Bigalke H, Krampfl K, Dengler R, Kollewe K (2009) A long-term follow-up of botulinum toxin A in cervical dystonia. Neurol Res 31:463–466
11. Truong D, Brodsky M, Lew M, Brashear A, Jankovic J, Molho E et al (2010) Long-term efficacy and safety of botulinum toxin type A (Dysport) in cervical dystonia. Parkinsonism Relat Disord 16:316–323
12. Dressler D, Eleopra R (2006) Clinical use of non-A botulinum toxins: botulinum toxin type B. Neurotox Res 9:121–125
13. Sanger TD, Kukke SN, Sherman-Levine S (2007) Botulinum toxin type B improves the speed of reaching in children with cerebral palsy and arm dystonia: an open-label, dose-escalation pilot study. J Child Neurol 22:116–122
14. Hallett M, Daroff RB (1996) Blepharospasm: report of a workshop. Neurology 46:1213–1218
15. Defazio G, Livrea P (2002) Epidemiology of primary blepharospasm. Mov Disord 17:7–12
16. The Epidemiological Study of Dystonia in Europe (ESDE) Collaborative Group (1999) Sex-related influences on the frequency and age of onset of primary dystonia. Epidemiologic Study of Dystonia in Europe (ESDE) Collaborative Group. Neurology 53:1871–1873
17. Fahn S (1988) Blepharospasm: a form of focal dystonia. Adv Neurol 49:125–133
18. Hallett M, Evinger C, Jankovic J, Stacy M (2008) Update on blepharospasm: report from the BEBRF International Workshop. Neurology 71:1275–1282
19. Weiss EM, Hershey T, Karimi M, Racette B, Tabbal SD, Mink JW et al (2006) Relative risk of spread of symptoms among the focal onset primary dystonias. Mov Disord 21:1175–1181
20. Svetel M, Pekmezovic T, Jovic J, Ivanovic N, Dragasevic N, Maric J et al (2007) Spread of primary dystonia in relation to initially affected region. J Neurol 254:879–883
21. Abbruzzese G, Berardelli A, Girlanda P, Marchese R, Martino D, Morgante F et al (2008) Long-term assessment of the risk of spread in primary late-onset focal dystonia. J Neurol Neurosurg Psychiatry 79:392–396
22. Grandas F, Elston J, Quinn N, Marsden CD (1988) Blepharospasm: a review of 264 patients. J Neurol Neurosurg Psychiatry 51:767–772
23. Quartarone A, Sant'angelo A, Battaglia F, Bagnato S, Rizzo V, Morgante F et al (2006) Enhanced long-term potentiation-like plasticity of the trigeminal blink reflex circuit in blepharospasm. J Neurosci 26:716–721
24. Scott AB, Kennedy RA, Stubbs MA (1985) Botulinum toxin injections as a treatment for blepharospasm. Arch Ophthalmol 103:347–350
25. Balash Y, Giladi N (2004) Efficacy of pharmacological treatment of dystonia: evidence-based review including meta-analysis of the effect of botulinum toxin and other cure options. Eur J Neurol 11:361–370

26. Jost WH, Kohl A (2001) Botulinum toxin: evidence-based medicine criteria in blepharospasm and hemifacial spasm. J Neurol 248(Suppl 1):21–24
27. MacAndie K, Kemp E (2004) Impact on quality of life of botulinum toxin treatments for essential blepharospasm. Orbit 23:207–210
28. Aramideh M, Ongerboer de Visser BW, Brans JW, Koelman JH, Speelman JD (1995) Pretarsal application of botulinum toxin for treatment of blepharospasm. J Neurol Neurosurg Psychiatry 59:309–311
29. Albanese A, Bentivoglio AR, Colosimo C, Galardi G, Maderna L, Tonali P (1996) Pretarsal injections of botulinum toxin improve blepharospasm in previously unresponsive patients. J Neurol Neurosurg Psychiatry 60:693–694
30. Jankovic J (1996) Pretarsal injection of botulinum toxin for blepharospasm and apraxia of eyelid opening. J Neurol Neurosurg Psychiatry 60:704
31. Cakmur R, Ozturk V, Uzunel F, Donmez B, Idiman F (2002) Comparison of preseptal and pretarsal injections of botulinum toxin in the treatment of blepharospasm and hemifacial spasm. J Neurol 249:64–68
32. Levy RL, Berman D, Parikh M, Miller NR (2006) Supramaximal doses of botulinum toxin for refractory blepharospasm. Ophthalmology 113:1665–1668
33. Pang AL, O'Day J (2006) Use of high-dose botulinum A toxin in benign essential blepharospasm: is too high too much? Clin Experiment Ophthalmol 34:441–444
34. Hsiung GY, Das SK, Ranawaya R, Lafontaine AL, Suchowersky O (2002) Long-term efficacy of botulinum toxin A in treatment of various movement disorders over a 10-year period. Mov Disord 17:1288–1293
35. Sampaio C, Ferreira JJ, Simoes F, Rosas MJ, Magalhaes M, Correia AP et al (1997) DYSBOT: a single-blind, randomized parallel study to determine whether any differences can be detected in the efficacy and tolerability of two formulations of botulinum toxin type A—Dysport and Botox—assuming a ratio of 4:1. Mov Disord 12:1013–1018
36. Nussgens Z, Roggenkamper P (1997) Comparison of two botulinum-toxin preparations in the treatment of essential blepharospasm. Graefes Arch Clin Exp Ophthalmol 235:197–199
37. Bihari K (2005) Safety, effectiveness, and duration of effect of BOTOX after switching from Dysport for blepharospasm, cervical dystonia, and hemifacial spasm dystonia, and hemifacial spasm. Curr Med Res Opin 21:433–438
38. Badarny S, Susel Z, Honigman S (2008) Effectivity of Dysport in patients with blepharospasm and hemifacial spasm who experienced failure with Botox. Isr Med Assoc J 10:520–522
39. Roggenkamper P, Jost WH, Bihari K, Comes G, Grafe S (2006) Efficacy and safety of a new botulinum toxin type A free of complexing proteins in the treatment of blepharospasm. J Neural Transm 113:303–312
40. Wabbels B, Reichel G, Fulford-Smith A, Wright N, Roggenkamper P (2011) Double-blind, randomised, parallel group pilot study comparing two botulinum toxin type A products for the treatment of blepharospasm. J Neural Transm 118:233–239
41. Jankovic J (2009) Clinical efficacy and tolerability of Xeomin in the treatment of blepharospasm. Eur J Neurol 16(Suppl 2):14–18
42. Rieder CR, Schestatsky P, Socal MP, Monte TL, Fricke D, Costa J et al (2007) A double-blind, randomized, crossover study of Prosigne versus Botox in patients with blepharospasm and hemifacial spasm. Clin Neuropharmacol 30:39–42
43. Quagliato EM, Carelli EF, Viana MA (2010) Prospective, randomized, double-blind study, comparing botulinum toxins type a botox and prosigne for blepharospasm and hemifacial spasm treatment. Clin Neuropharmacol 33:27–31
44. Colosimo C, Chianese M, Giovannelli M, Contarino MF, Bentivoglio AR (2003) Botulinum toxin type B in blepharospasm and hemifacial spasm. J Neurol Neurosurg Psychiatry 74:687
45. Dutton JJ, White JJ, Richard MJ (2006) Myobloc for the treatment of benign essential blepharospasm in patients refractory to Botox. Ophthal Plast Reconstr Surg 22:173–177
46. Price J, Farish S, Taylor H, O'Day J (1997) Blepharospasm and hemifacial spasm. Randomized trial to determine the most appropriate location for botulinum toxin injections. Ophthalmology 104:865–868

47. Nutt JG, Muenter MD, Aronson A, Kurland LT, Melton LJ (1988) Epidemiology of focal and generalyzed dystonia in Rochester, Minnesota. Mov Disord 3:188–194

48. Defazio G, Abbruzzese G, Livrea P, Berardelli A (2004) Epidemiology of primary dystonia. Lancet Neurol 3:673–678

49. Jankovic J, Orman J (1987) Botulinum A toxin for cranial-cervical dystonia: a double-blind, placebo-controlled study. Neurology 37:616–623

50. Chan J, Brin MF, Fahn S (1991) Idiopathic cervical dystonia: clinical characteristics. Mov Disord 6:119–126

51. Camfield L, Ben-Shlomo Y, Warner TT (2002) Impact of cervical dystonia on quality of life. Mov Disord 17:838–841

52. Hagenah JM, Vieregge A, Vieregge P (2001) Radiculopathy and myelopathy in patients with primary cervical dystonia. Eur Neurol 45:236–240

53. Hung SW, Hamani C, Lozano AM, Poon YY, Piboolnurak P, Miyasaki JM et al (2007) Long-term outcome of bilateral pallidal deep brain stimulation for primary cervical dystonia. Neurology 68:457–459

54. Cacciola F, Farah JO, Eldridge PR, Byrne P, Varma TK (2010) Bilateral deep brain stimulation for cervical dystonia: long-term outcome in a series of 10 patients. Neurosurgery 67:957–963

55. Jankovic J (2006) Treatment of dystonia. Lancet Neurol 5:864–872

56. Molho E, Jankovic J, Lew M (2008) Role of botulinum toxin in the treatment of cervical dystonia. Neurol Clin 26(Suppl 1):43–53

57. Tsui JK, Eisen A, Mak E, Carruthers J, Scott A, Calne DB (1985) A pilot study on the use of botulinum toxin in spasmodic torticollis. Can J Neurol Sci 12:314–316

58. Brans JW, Lindeboom R, Snoek JW, Zwarts MJ, van Weerden TW, Brunt ER et al (1996) Botulinum toxin versus trihexyphenidyl in cervical dystonia: a prospective, randomized, double-blind controlled trial. Neurology 46:1066–1072

59. Sampaio C, Costa J, Ferreira JJ (2004) Clinical comparability of marketed formulations of botulinum toxin. Mov Disord 19(Suppl 8):S129–S136

60. Odergren T, Hjaltason H, Kaakkola S, Solders G, Hanko J, Fehling C et al (1998) A double blind, randomised, parallel group study to investigate the dose equivalence of Dysport and Botox in the treatment of cervical dystonia. J Neurol Neurosurg Psychiatry 64:6–12

61. Ranoux D, Gury C, Fondarai J, Mas JL, Zuber M (2002) Respective potencies of Botox and Dysport: a double blind, randomised, crossover study in cervical dystonia. J Neurol Neurosurg Psychiatry 72:459–462

62. Marchetti A, Magar R, Findley L, Larsen JP, Pirtosek Z, Ruzicka E et al (2005) Retrospective evaluation of the dose of Dysport and BOTOX in the management of cervical dystonia and blepharospasm: the REAL DOSE study. Mov Disord 20:937–944

63. Lu CS, Chen RS, Tsai CH (1995) Double-blind, placebo-controlled study of botulinum toxin injections in the treatment of cervical dystonia. Journal of the Formosan Medical Association 94(4):189–192

64. Wissel J, Kanovsky P, Ruzicka E, Bares M, Hortova H, Streitova H et al (2001) Efficacy and safety of a standardised 500 unit dose of Dysport (clostridium botulinum toxin type A haemaglutinin complex) in a heterogeneous cervical dystonia population: results of a prospective, multicentre, randomised, double-blind, placebo-controlled, parallel group study. J Neurol 248:1073–1078

65. Poewe W, Deuschl G, Nebe A, Feifel E, Wissel J, Benecke R et al (1998) What is the optimal dose of botulinum toxin A in the treatment of cervical dystonia? Results of a double blind, placebo controlled, dose ranging study using Dysport. German Dystonia Study Group. J Neurol Neurosurg Psychiatry 64:13–17

66. Benecke R, Jost WH, Kanovsky P, Ruzicka E, Comes G, Grafe S (2005) A new botulinum toxin type A free of complexing proteins for treatment of cervical dystonia. Neurology 64:1949–1951

67. Pappert EJ, Germanson T (2008) Botulinum toxin type B vs. type A in toxin-naive patients with cervical dystonia: Randomized, double-blind, noninferiority trial. Mov Disord 23:510–517

68. Chen R, Karp BI, Hallett M (1998) Botulinum toxin type F for treatment of dystonia: long-term experience. Neurology 51:1494–1496
69. Houser MK, Sheean GL, Lees AJ (1998) Further studies using higher doses of botulinum toxin type F for torticollis resistant to botulinum toxin type A. J Neurol Neurosurg Psychiatry 64:577–580
70. Costa J, Espirito-Santo C, Borges A, Ferreira J, Coelho M, Moore P et al (2005) Botulinum toxin type A therapy for cervical dystonia. Cochrane Database Syst Rev CD003633
71. Brashear A, Watts MW, Marchetti A, Magar R, Lau H, Wang L (2000) Duration of effect of botulinum toxin type A in adult patients with cervical dystonia: a retrospective chart review. Clin Ther 22:1516–1524
72. Kessler KR, Skutta M, Benecke R (1999) Long-term treatment of cervical dystonia with botulinum toxin A: efficacy, safety, and antibody frequency. German Dystonia Study Group. J Neurol 246:265–274
73. Costa J, Espirito-Santo C, Borges A, Ferreira J, Coelho M, Moore P et al (2005) Botulinum toxin type B for cervical dystonia. Cochrane Database Syst Rev CD004315
74. Chapman MA, Barron R, Tanis DC, Gill CE, Charles PD (2007) Comparison of botulinum neurotoxin preparations for the treatment of cervical dystonia. Clin Ther 29:1325–1337
75. Tintner R, Gross R, Winzer UF, Smalky KA, Jankovic J (2005) Autonomic function after botulinum toxin type A or B: a double-blind, randomized trial. Neurology 65:765–767
76. Jankovic J, Schwartz KS (1991) Clinical correlates of response to botulinum toxin injections. Arch Neurol 48:1253–1256
77. Brin MF, Lew MF, Adler CH, Comella CL, Factor SA, Jankovic J et al (1999) Safety and efficacy of NeuroBloc (botulinum toxin type B) in type A- resistant cervical dystonia. Neurology 53:1431–1438
78. Tan EK, Jankovic J (1999) Botulinum toxin A in patients with oromandibular dystonia: long-term follow-up. Neurology 53:2102–2107
79. Ilic TV, Potter M, Holler I, Deuschl G, Volkmann J (2005) Praying-induced oromandibular dystonia. Mov Disord 20:385–386
80. Bhatia KP, Bhatt MH, Marsden CD (1993) The causalgia-dystonia syndrome. Brain 116:843–851
81. Tan EK, Chan LL, Wong MC (2003) Levodopa-induced oromandibular dystonia in progressive supranuclear palsy. Clin Neurol Neurosurg 105:132–134
82. Klawans HL, Tanner CM (1988) Cholinergic pharmacology of blepharospasm with oromandibular dystonia (Meige's syndrome). Adv Neurol 49:443–450
83. Ondo WG, Hanna PA, Jankovic J (1999) Tetrabenazine treatment for tardive dyskinesia: assessment by randomized videotape protocol. Am J Psychiatry 156:1279–1281
84. Kenney C, Hunter C, Jankovic J (2007) Long-term tolerability of tetrabenazine in the treatment of hyperkinetic movement disorders. Mov Disord 22:193–197
85. Yoshida K, Kaji R, Kubori T, Kohara N, Iizuka T, Kimura J (1998) Muscle afferent block for the treatment of oromandibular dystonia. Mov Disord 13:699–705
86. Capelle HH, Weigel R, Krauss JK (2003) Bilateral pallidal stimulation for blepharospasm-oromandibular dystonia (Meige syndrome). Neurology 60:2017–2018
87. Romito LM, Elia AE, Franzini A, Bugiani O, Albanese A (2010) Low-voltage bilateral pallidal stimulation for severe Meige syndrome in a patient with primary segmental dystonia: case report. Neurosurgery 67(3 Suppl Operative):E308-1-E308/5
88. Blitzer A, Brin MF, Greene PE, Fahn S (1989) Botulinum toxin injection for the treatment of oromandibular dystonia. Ann Otol Rhinol Laryngol 98:93–97
89. Hermanowicz N, Truong DD (1991) Treatment of oromandibular dystonia with botulinum toxin. Laryngoscope 101:1216–1218
90. Cardoso F (2003) Toxina botunílica tipo B no manejo de distonia não-responsiva a toxina botunílica tipo A. Arq Neuropsiquiatr 61:607–610
91. Wan XH, Vuong KD, Jankovic J (2005) Clinical application of botulinum toxin type B in movement disorders and autonomic symptoms. Chin Med Sci J 20:44–47

92. Bhidayasiri R, Cardoso F, Truong DD (2006) Botulinum toxin in blepharospasm and oromandibular dystonia: comparing different botulinum toxin preparations. Eur J Neurol 13(Suppl 1):21–29

93. Hallett M, Benecke R, Blitzer A, Comella CL (2009) Treatment of focal dystonias with botulinum neurotoxin. Toxicon 54:628–633

94. Esper CD, Freeman A, Factor SA (2010) Lingual protrusion dystonia: frequency, etiology and botulinum toxin therapy. Parkinsonism Relat Disord 16:438–441

95. Blitzer A, Brin MF, Fahn S, Lovelace RE (1988) Clinical and laboratory characteristics of focal laryngeal dystonia: study of 110 cases. Laryngoscope 98:636–640

96. Adler CH, Edwards BW, Bansberg SF (1997) Female predominance in spasmodic dysphonia. J Neurol Neurosurg Psychiatry 63:688

97. Schweinfurth JM, Billante M, Courey MS (2002) Risk factors and demographics in patients with spasmodic dysphonia. Laryngoscope 112:220–223

98. Murry T, Woodson GE (1995) Combined-modality treatment of adductor spasmodic dysphonia with botulinum toxin and voice therapy. J Voice 9:460–465

99. Blitzer A, Brin MF, Fahn S, Lovelace RE (1988) Localized injections of botulinum toxin for the treatment of focal laryngeal dystonia (spastic dysphonia). Laryngoscope 98:193–197

100. Blitzer A, Brin MF, Stewart C, Aviv JE, Fahn S (1992) Abductor laryngeal dystonia: a series treated with botulinum toxin. Laryngoscope 102:163–167

101. Albanese A, Barnes MP, Bhatia KP, Fernandez-Alvarez E, Filippini G, Gasser T et al (2006) A systematic review on the diagnosis and treatment of primary (idiopathic) dystonia and dystonia plus syndromes: report of an EFNS/MDS-ES Task Force. Eur J Neurol 13:433–44

102. Ludlow CL, Naunton RF, Sedory SE, Schulz GM, Hallett M (1988) Effects of botulinum toxin injections on speech in adductor spasmodic dysphonia. Neurology 38:1220–1225

103. Bielamowicz S, Stager SV, Badillo A, Godlewski A (2002) Unilateral versus bilateral injections of botulinum toxin in patients with adductor spasmodic dysphonia. J Voice 16: 117–123

104. Blitzer A, Brin MF, Stewart CF (1998) Botulinum toxin management of spasmodic dysphonia (laryngeal dystonia): a 12-year experience in more than 900 patients. Laryngoscope 108:1435–1441

105. Elmiyeh B, Prasad VM, Upile T, Saunders N, Youl BD, Epstein R et al (2010) A single-centre retrospective review of unilateral and bilateral Dysport injections in adductor spasmodic dysphonia. Logoped Phoniatr Vocol 35:39–44

106. Upile T, Elmiyeh B, Jerjes W, Prasad V, Kafas P, Abiola J et al (2009) Unilateral versus bilateral thyroarytenoid Botulinum toxin injections in adductor spasmodic dysphonia: a prospective study. Head Face Med 5:20

107. Brin MF, Blitzer A, Fahn S, Gould W, Lovelace RE (1989) Adductor laryngeal dystonia (spastic dysphonia): treatment with local injections of botulinum toxin (Botox). Mov Disord 4:287–296

108. Whurr R, Nye C, Lorch M (1998) Meta-analysis of botulinum toxin treatment of spasmodic dysphonia: a review of 22 studies. Int J Lang Commun Disord 33(Suppl):327–329

109. Jankovic J, Schwartz K, Donovan DT (1990) Botulinum toxin treatment of cranial-cervical dystonia, spasmodic dysphonia, other focal dystonias and hemifacial spasm. J Neurol Neurosurg Psychiatry 53:633–639

110. Blitzer A, Sulica L (2001) Botulinum toxin: basic science and clinical uses in otolaryngology. Laryngoscope 111:218–226

111. Truong DD, Bhidayasiri R (2006) Botulinum toxin therapy of laryngeal muscle hyperactivity syndromes: comparing different botulinum toxin preparations. Eur J Neurol 13(Suppl 1):36–41

112. Maloney AP, Morrison MD (1994) A comparison of the efficacy of unilateral versus bilateral botulinum toxin injections in the treatment of adductor spasmodic dysphonia. J Otolaryngol 23:160–164

113. Guntinas-Lichius O (2003) Injection of botulinum toxin type B for the treatment of otolaryngology patients with secondary treatment failure of botulinum toxin type A. Laryngoscope 113:743–745

114. Adler CH, Bansberg SF, Krein-Jones K, Hentz JG (2004) Safety and efficacy of botulinum toxin type B (Myobloc) in adductor spasmodic dysphonia. Mov Disord 19:1075–1079
115. Blitzer A (2005) Botulinum toxin A and B: a comparative dosing study for spasmodic dysphonia. Otolaryngol Head Neck Surg 133:836–838
116. Bielamowicz S, Squire S, Bidus K, Ludlow CL (2001) Assessment of posterior cricoarytenoid botulinum toxin injections in patients with abductor spasmodic dysphonia. Ann Otol Rhinol Laryngol 110:406–412
117. Meleca RJ, Hogikyan ND, Bastian RW (1997) A comparison of methods of botulinum toxin injection for abductory spasmodic dysphonia. Otolaryngol Head Neck Surg 117: 487–492
118. Stong BC, DelGaudio JM, Hapner ER, Johns MM, III (2005) Safety of simultaneous bilateral botulinum toxin injections for abductor spasmodic dysphonia. Arch Otolaryngol Head Neck Surg 131:793–795
119. Blitzer A (2010) Spasmodic dysphonia and botulinum toxin: experience from the largest treatment series. Eur J Neurol 17(Suppl 1):28–30
120. Karp BI, Cole RA, Cohen LG, Grill S, Lou JS, Hallett M (1994) Long-term botulinum toxin treatment of focal hand dystonia. Neurology 44:70–76
121. Karp BI (2004) Botulinum toxin treatment of occupational and focal hand dystonia. Mov Disord 19(Suppl 8):S116–S119
122. Kruisdijk JJ, Koelman JH, Ongerboer de Visser BW, De Haan RJ, Speelman JD (2007) Botulinum toxin for writer's cramp: a randomised, placebo-controlled trial and 1-year follow-up. J Neurol Neurosurg Psychiatry 78:264–270
123. Yoshimura DM, Aminoff MJ, Olney RK (1992) Botulinum toxin therapy for limb dystonias. Neurology 42:627–630
124. Tsui JK, Bhatt M, Calne S, Calne DB (1993) Botulinum toxin in the treatment of writer's cramp: a double-blind study. Neurology 43:183–185
125. Das CP, Dressler D, Hallett M (2006) Botulinum toxin therapy of writer's cramp. Eur J Neurol 13(Suppl 1):55–59
126. Molloy FM, Shill HA, Kaelin-Lang A, Karp BI (2002) Accuracy of muscle localization without EMG: implications for treatment of limb dystonia. Neurology 58: 805–807
127. Duarte J, Sempere AP, Coria F, Claveria LE, Frech FA, Mataix AL et al (1995) Isolated idiopathic adult-onset foot dystonia and treatment with botulinum toxin. J Neurol 242: 114–115
128. Schneider SA, Edwards MJ, Grill SE, Goldstein S, Kanchana S, Quinn NP et al (2006) Adult-onset primary lower limb dystonia. Mov Disord 21:767–771
129. Singer C, Papapetropoulos S (2006) Adult-onset primary focal foot dystonia. Parkinsonism Relat Disord 12:57–60
130. Wang A, Jankovic J (1998) Hemifacial spasm: clinical findings and treatment. Muscle Nerve 21:1740–1747
131. Micheli F, Scorticati MC, Gatto E, Cersosimo G, Adi J (1994) Familial hemifacial spasm. Mov Disord 9:330–332
132. Auger RG, Whisnant JP (1990) Hemifacial spasm in Rochester and Olmsted County, Minnesota, 1960 to 1984. Arch Neurol 47:1233–1234
133. Tan EK, Chan LL, Lim SH, Lim WE, Khoo JB, Tan KP (1999) Role of magnetic resonance imaging and magnetic resonance angiography in patients with hemifacial spasm. Ann Acad Med Singapore 28:169–173
134. Martinelli P, Giuliani S, Ippoliti M (1992) Hemifacial spasm due to peripheral injury of facial nerve: a nuclear syndrome? Mov Disord 7:181–184
135. Holds JB, Anderson RL, Jordan DR, Patrinely JR (1990) Bilateral hemifacial spasm. J Clin Neuroophthalmol 10:153–154
136. Tan EK, Jankovic J (1999) Bilateral hemifacial spasm: a report of five cases and a literature review. Mov Disord 14:345–349

137. Alexander GE, Moses H (1982) Carbamazepine for hemifacial spasm. Neurology 32: 286–287
138. Kemp LW, Reich SG (2004) Hemifacial spasm. Curr Treat Options Neurol 6:175–179
139. Chung SS, Chang JH, Choi JY, Chang JW, Park YG (2001) Microvascular decompression for hemifacial spasm: a long-term follow-up of 1,169 consecutive cases. Stereotact Funct Neurosurg 77:190–193
140. Samii M, Gunther T, Iaconetta G, Muehling M, Vorkapic P, Samii A (2002) Microvascular decompression to treat hemifacial spasm: long-term results for a consecutive series of 143 patients. Neurosurgery 50:712–718
141. Loeser JD, Chen J (1983) Hemifacial spasm: treatment by microsurgical facial nerve decompression. Neurosurgery 13:141–146
142. Piatt JH Jr, Wilkins RH (1984) Treatment of tic douloureux and hemifacial spasm by posterior fossa exploration: therapeutic implications of various neurovascular relationships. Neurosurgery 14:462–471
143. Ito M, Hasegawa M, Hoshida S, Miwa T, Furukawa M (2004) Successful treatment of hemifacial spasm with selective facial nerve block using doxorubicin (adriamycin) under local anesthesia. Acta Otolaryngol 124:217–220
144. Elston JS (1986) Botulinum toxin treatment of hemifacial spasm. J Neurol Neurosurg Psychiatry 49:827–829
145. Taylor JD, Kraft SP, Kazdan MS, Flanders M, Cadera W, Orton RB (1991) Treatment of blepharospasm and hemifacial spasm with botulinum A toxin: a Canadian multicentre study. Can J Ophthalmol 26:133–138
146. Elston JS (1992) The management of blepharospasm and hemifacial spasm. J Neurol 239:5–8
147. Flanders M, Chin D, Boghen D (1993) Botulinum toxin: preferred treatment for hemifacial spasm. Eur Neurol 33:316–319
148. Mauriello JA, Leone T, Dhillon S, Pakeman B, Mostafavi R, Yepez MC (1996) Treatment choices of 119 patients with hemifacial spasm over 11 years. Clin Neurol Neurosurg 98: 213–216
149. Jitpimolmard S, Tiamkao S, Laopaiboon M (1998) Long term results of botulinum toxin type A (Dysport) in the treatment of hemifacial spasm: a report of 175 cases. J Neurol Neurosurg Psychiatry 64:751–757
150. Defazio G, Abbruzzese G, Girlanda P, Vacca L, Curra A, De Salvia R et al (2002) Botulinum toxin A treatment for primary hemifacial spasm: a 10-year multicenter study. Arch Neurol 59:418–420
151. Park YC, Lim JK, Lee DK, Yi SD (1993) Botulinum A toxin treatment of hemifacial spasm and blepharospasm. J Korean Med Sci 8:334–340
152. Yoshimura DM, Aminoff MJ, Tami TA, Scott AB (1992) Treatment of hemifacial spasm with botulinum toxin. Muscle Nerve 15:1045–1049
153. Costa J, Espirito-Santo C, Borges A, Ferreira JJ, Coelho M, Moore P et al (2005) Botulinum toxin type A therapy for hemifacial spasm. Cochrane Database Syst Rev (1):CD004899
154. Bentivoglio AR, Fasano A, Ialongo T, Soleti F, Lo FS, Albanese A (2009) Outcome predictors, efficacy and safety of Botox and Dysport in the long-term treatment of hemifacial spasm. Eur J Neurol 16:392–398
155. Trosch RM, Adler CH, Pappert EJ (2007) Botulinum toxin type B (Myobloc) in subjects with hemifacial spasm: results from an open-label, dose-escalation safety study. Mov Disord 22:1258–1264
156. Dressler D (2009) Routine use of Xeomin in patients previously treated with Botox: long term results. Eur J Neurol 16(Suppl 2):2–5
157. Eleopra R, Tugnoli V, Rossetto O, Montecucco C, De Grandis D (1997) Botulinum neurotoxin serotype C: a novel effective botulinum toxin therapy in human. Neurosci Lett 224:91–94
158. Chen RS, Lu CS, Tsai CH (1996) Botulinum toxin A injection in the treatment of hemifacial spasm. Acta Neurol Scand 94:207–211
159. Yu YL, Fong KY, Chang CM (1992) Treatment of idiopathic hemifacial spasm with botulinum toxin. Acta Neurol Scand 85:55–57

160. Ward AB (2002) A summary of spasticity management-a treatment algorithm. Eur J Neurol 9(Suppl. 1):48–52
161. Brashear A, Gordon MF, Elovic E, Kassicieh VD, Marciniak C, Do M et al (2002) Intramuscular injection of botulinum toxin for the treatment of wrist and finger spasticity after a stroke. N Engl J Med 347:395–400
162. Ward AB, Aguilar M, De Beyl Z, Gedin S, Kanovsky P, Molteni F et al (2003) Use of botulinum toxin type A in management of adult spasticity-a European consensus statement. J Rehabil Med 35:98–99
163. Pathak MS, Nguyen HT, Graham HK, Moore AP (2006) Management of spasticity in adults: practical application of botulinum toxin. Eur J Neurol 13(Suppl 1):42–50
164. Koman LA, Paterson SB, Balkrishnan R (2003) Spasticity associated with cerebral palsy in children: guidelines for the use of botulinum A toxin. Paediatr Drugs 5:11–23
165. Bakheit AM, Bower E, Cosgrove A, Fox M, Morton R, Phillips S et al (2001) Opinion statement on the minimal acceptable standards of healthcare in cerebral palsy. Disabil Rehabil 23:578–582
166. Schwerin A, Berweck S, Fietzek UM, Heinen F (2004) Botulinum toxin B treatment in children with spastic movement disorders: a pilot study. Pediatr Neurol 31:109–113
167. Francisco GE (2004) Botulinum toxin: dosing and dilution. Am J Phys Med Rehabil 83(Suppl):S30–S37
168. Elia AE, Filippini G, Calandrella D, Albanese A (2009) Botulinum neurotoxins for post-stroke spasticity in adults: a systematic review. Mov Disord 24:801–812
169. Gracies JM, Lugassy M, Weisz DJ, Vecchio M, Flanagan S, Simpson DM (2009) Botulinum toxin dilution and endplate targeting in spasticity: a double-blind controlled study. Arch Phys Med Rehabil 90:9–16
170. Simpson DM, Gracies JM, Graham HK, Miyasaki JM, Naumann M, Russman B et al (2008) Botulinum neurotoxin for the treatment of spasticity (an evidence-based review): report of the Therapeutics and Technology Assessment Subcommittee of the American Academy of Neurology. Neurology 70:1691–1698
171. Turkel CC, Bowen B, Liu J, Brin MF (2006) Pooled analysis of the safety of botulinum toxin type A in the treatment of poststroke spasticity. Arch Phys Med Rehabil 87:786–792
172. Brashear A, McAfee AL, Kuhn ER, Fyffe J (2004) Botulinum toxin type B in upper-limb poststroke spasticity: a double-blind, placebo-controlled trial. Arch Phys Med Rehabil 85:705–709
173. Rosales RL, Chua-Yap AS (2008) Evidence-based systematic review on the efficacy and safety of botulinum toxin-A therapy in post-stroke spasticity. J Neural Transm 115:617–623
174. Kaji R, Osako Y, Suyama K, Maeda T, Uechi Y, Iwasaki M (2010) Botulinum toxin type A in post-stroke lower limb spasticity: a multicenter, double-blind, placebo-controlled trial. J Neurol 257:1330–1337
175. Shakespeare DT, Boggild M, Young C (2003) Anti-spasticity agents for multiple sclerosis. Cochrane Database Syst Rev (4):CD001332
176. Snow BJ, Tsui JK, Bhatt MH (1990) Treatment of spasticity with botulinum toxin: a double-blind study. Ann Neurol 28:512–515
177. Hyman N, Barnes M, Bhakta B, Cozens A, Bakheit M, Kreczy-Kleedorfer B et al (2000) Botulinum toxin (Dysport) treatment of hip adductor spasticity in multiple sclerosis: a prospective, randomised, double blind, placebo controlled, dose ranging study. J Neurol Neurosurg Psychiatry 68:707–712
178. Restivo DA, Tinazzi M, Patti F, Palmeri A, Maimone D (2003) Botulinum toxin treatment of painful tonic spasms in multiple sclerosis. Neurology 61:719–720
179. Giovannelli M, Borriello G, Castri P, Prosperini L, Pozzilli C (2007) Early physiotherapy after injection of botulinum toxin increases the beneficial effects on spasticity in patients with multiple sclerosis. Clin Rehabil 21:331–337
180. Ward AB (2003) Long-term modification of spasticity. J Rehabil Med 41(Suppl):60–65

181. Koman LA, Mooney JF, Smith B, Goodman A, Mulvaney T (1993) Management of cerebral palsy with botulinum A toxin: preliminary investigation. J Pediatr Orthoped 13:489–495
182. Lukban MB, Rosales RL, Dressler D (2009) Effectiveness of botulinum toxin A for upper and lower limb spasticity in children with cerebral palsy: a summary of evidence. J Neural Transm 116:319–331
183. Kuehn BM (2009) FDA requires black box warnings on labeling for botulinum toxin products. JAMA 301:2316
184. Deuschl G, Bain P, Brin M (1998) Consensus statement of the movement disorder society on tremor. Ad Hoc Scientific Committee. Mov Disord 13(Suppl. 3):2–23
185. Trosch RM, Pullman SL (1994) Botulinum toxin A injections for the treatment of hand tremors. Mov Disord 9:601–609
186. Henderson JM, Ghika JA, Van Melle G, Haller E, Einstein R (1996) Botulinum toxin A in non-dystonic tremors. Eur Neurol 36(1):29–35
187. Koller WC, Hristova A, Brin M (2000) Pharmacologic treatment of essential tremor. Neurology 54(Suppl 4):S30–S38
188. Brin MF, Lyons KE, Doucette J, Adler CH, Caviness JN, Comella CL et al (2001) A randomized, double masked, controlled trial of botulinum toxin type A in essential hand tremor. Neurology 56:1523–1528
189. Jankovic J, Schwartz K (1991) Botulinum toxin treatment of tremors. Neurology 41: 1185–1188
190. Pullman S, Greene PF, Fahn S (1996) Approach to the treatment of limb disorders with botulinum toxin A. Experience with 187 patients. Arch Neurol 53:617–624
191. Jankovic J, Schwartz K, Clemence W, Aswad A, Mordaunt J (1996) A randomized, double-blind, placebo-controlled study to evaluate botulinum toxin type A in essential hand tremor. Mov Disord 11:250–256
192. Papapetropoulos S, Singer C (2006) Treatment of primary writing tremor with botulinum toxin type a injections: report of a case series. Clin Neuropharmacol 29:364–367
193. Wissel J, Masuhr F, Schelosky L, Ebersbach G, Poewe W (1997) Quantitative assessment of botulinum toxin treatment in 43 patients with head tremor. Mov Disord 12(5):722–726
194. Hertegard S, Granqvist S, Lindestad PA (2000) Botulinum toxin injections for essential voice tremor. Ann Otol Rhinol Laryngol 109:204–209
195. Warrick P, Dromey C, Irish JC, Durkin L, Pakiam A, Lang A (2000) Botulinum toxin for essential tremor of the voice with multiple anatomical sites of tremor: a crossover design study of unilateral versus bilateral injection. Laryngoscope 110:1366–1374
196. Pahwa R, Busenbark K, Swanson-Hyland EF, Dubinsky RM, Hubble JP, Gray C et al (1995) Botulinum toxin treatment of essential head tremor. Neurology 45:822–824
197. Adler CH, Bansberg SF, Hentz JG, Ramig LO, Buder EH, Witt K et al (2004) Botulinum toxin type A for treating voice tremor. Arch Neurol 61:1416–1420
198. Schneider SA, Edwards MJ, Cordivari C, Macleod WN, Bhatia KP (2006) Botulinum toxin A may be efficacious as treatment for jaw tremor in Parkinson's disease. Mov Disord 21: 1722–1724
199. Gonzalez-Alegre P, Kelkar P, Rodnitzky RL (2006) Isolated high-frequency jaw tremor relieved by botulinum toxin injections. Mov Disord 21:1049–1050
200. Deuschl G, Lohle E, Heinen F, Lucking C (1991) Ear click in palatal tremor: its origin and treatment with botulinum toxin. Neurology 41:1677–1679
201. Penney SE, Bruce IA, Saeed SR (2006) Botulinum toxin is effective and safe for palatal tremor: a report of five cases and a review of the literature. J Neurol 253:857–860
202. Jankovic J (1994) Botulinum toxin in the treatment of dystonic tics. Mov Disord 9: 347–349
203. Scott BL, Jankovic J, Donovan DT (1996) Botulinum toxin injection into vocal cord in the treatment of malignant coprolalia associated with Tourette's syndrome. Mov Disord 11: 431–433
204. Trimble MR, Whurr R, Brookes G, Robertson MM (1998) Vocal tics in Gilles de la Tourette syndrome treated with botulinum toxin injections. Mov Disord 13:617–619

205. Krauss JK, Jankovic J (1996) Severe motor tics causing cervical myelopathy in Tourette's syndrome. Mov Disord 11:563–566
206. Kwak CH, Hanna PA, Jankovic J (2000) Botulinum toxin in the treatment of tics. Arch Neurol 57:1190–1193
207. Rath JJ, Tavy DL, Wertenbroek AA, Van Woerkom TC, de Bruijn SF (2010) Botulinum toxin type A in simple motor tics: short-term and long-term treatment-effects. Parkinsonism Relat Disord 16:478–481
208. Marras C, Andrews D, Sime E, Lang AE (2001) Botulinum toxin for simple motor tics: a randomized, double-blind, controlled clinical trial. Neurology 56:605–610
209. Porta M, Maggioni G, Ottaviani F, Schindler A (2004) Treatment of phonic tics in patients with Tourette's syndrome using botulinum toxin type A. Neurol Sci 24:420–423
210. Cheung MY, Shahed J, Jankovic J (2007) Malignant Tourette syndrome. Mov Disord 22:1743–1750
211. Aguirregomozcorta M, Ramio-Torrenta L, Gich J, Quiles A, Genis D (2008) Paroxysmal dystonia and pathological laughter as a first manifestation of multiple sclerosis. Mult Scler 14:262–265
212. Truong DD, Hermanowicz N, Rontal M (1990) Botulinum toxin in treatment of tardive dyskinetic syndrome. J Clin Psychopharmacol 10:438–439
213. Stip E, Faughnan M, Desjardin I, Labrecque R (1992) Botulinum toxin in a case of severe tardive dyskinesia mixed with dystonia. Br J Psychiatry 161:867–868
214. Kaufman DM (1994) Use of botulinum toxin injections for spasmodic torticollis of tardive dystonia. J Neuropsychiatry Clin Neurosci 6:50–53
215. Yasufuku-Takano J, Sakurai M, Kanazawa I, Nagaoka M (1995) Successful treatment of intractable tardive dyskinesia with botulinum toxin. J Neurol Neurosurg Psychiatry 58:511–512
216. Van Zandijcke M, Marchau MM (1990) Treatment of bruxism with botulinum toxin injections. J Neurol Neurosurg Psychiatry 53:530
217. Diamond A, Shahed J, Azher S, Dat-Vuong K, Jankovic J (2006) Globus pallidus deep brain stimulation in dystonia. Mov Disord 21:692–695
218. Albanese A, Bentivoglio AR (2007) Botulinum toxin in movement disorders. In: Jankovic J, Tolosa E (eds) Parkinson's disease and movement disorders, 5th edn. Lippincott Williams & Wilkins, Philadelphia, pp 605–619
219. Jankovic J, Tintner R (2001) Dystonia and parkinsonism. Parkinsonism Relat Disord 8:109–121
220. Jankovic J, Stacy M (2007) Medical management of levodopa-associated motor complications in patients with Parkinson's disease. CNS Drugs 21:677–692
221. Pacchetti C, Albani G, Martignoni E, Godi L, Alfonsi E, Nappi G (1995) "Off" painful dystonia in Parkinson's disease treated with botulinum toxin. Mov Disord 10:333–336
222. Ashour R, Jankovic J (2006) Joint and skeletal deformities in Parkinson's disease, multiple system atrophy, and progressive supranuclear palsy. Mov Disord 21:1856–1863
223. Papapetropoulos S, Baez S, Zitser J, Sengun C, Singer C (2008) Retrocollis: classification, clinical phenotype, treatment outcomes and risk factors. Eur Neurol 59:71–75
224. Ghika J, Nater B, Henderson J, Bogousslavsky J, Regli F (1997) Delayed segmental axial dystonia of the trunk on standing after lumbar disk operation. J Neurol Sci 152(2):193–197
225. Comella CL, Shannon KM, Jaglin J (1998) Extensor truncal dystonia: successful treatment with botulinum toxin injections. Mov Disord 13:552–555
226. Bonanni L, Thomas A, Varanese S, Scorrano V, Onofrj M (2007) Botulinum toxin treatment of lateral axial dystonia in Parkinsonism. Mov Disord 22:2097–2103
227. Djaldetti R, Mosberg-Galili R, Sroka H, Merims D, Melamed E (1999) Camptocormia (bent spine) in patients with Parkinson's disease. Characterization and possible pathogenesis of an unusual phenomenon. Mov Disord 14:443–447
228. Azher SN, Jankovic J (2005) Camptocormia: pathogenesis, classification, and response to therapy. Neurology 65:355–359

229. Bloch F, Houeto JL, Tezenas du MS, Bonneville F, Etchepare F, Welter ML et al (2006) Parkinson's disease with camptocormia. J Neurol Neurosurg Psychiatry 77:1223–1228
230. von Coelln R, Raible A, Gasser T, Asmus F (2008) Ultrasound-guided injection of the iliopsoas muscle with botulinum toxin in camptocormia. Mov Disord 23:889–892
231. Ashour R, Tintner R, Jankovic J (2005) Striatal deformities of the hand and foot in Parkinson's disease. Lancet Neurol 4:423–431
232. Giladi N, Meer J, Honigman S (1994) The use of botulinum toxin to treat "striatal" toes. J Neurol Neurosurg Psychiatry 57:659
233. Cordivari C, Misra VP, Catania S, Lees AJ (2001) Treatment of dystonic clenched fist with botulinum toxin. Mov Disord 16:907–913
234. Jankovic J (2008) Parkinson's disease: clinical features and diagnosis. J Neurol Neurosurg Psychiatry 79:368–376
235. Giladi N (2001) Freezing of gait. Clinical overview. Adv Neurol 87:191–197
236. Giladi N, Honigman S (1997) Botulinum toxin injections to one leg alleviate freezing of gait in a patient with Parkinson's disease. Mov Disord 12:1085–1086
237. Giladi N, Gurevich T, Shabtai H, Paleacu D, Simon ES (2001) The effect of botulinum toxin injections to the calf muscles on freezing of gait in parkinsonism: a pilot study. J Neurol 248:572–576
238. Wieler M, Camicioli R, Jones CA, Martin WR (2005) Botulinum toxin injections do not improve freezing of gait in Parkinson disease. Neurology 65:626–628
239. Gurevich T, Peretz C, Moore O, Weizmann N, Giladi N (2007) The effect of injecting botulinum toxin type a into the calf muscles on freezing of gait in Parkinson's disease: a double blind placebo-controlled pilot study. Mov Disord 22:880–883
240. Blitzer A, Brin MF, Fahn S, Lange D, Lovelace RE (1986) Botulinum toxin (Botox) for the treatment of "spastic dysphonia" as part of a trial of toxin injections for the treatment of other cranial dystonias. Laryngoscope 96:1300–1301
241. Cohen LG, Hallett M, Geller BD, Hochberg F (1989) Treatment of focal dystonias of the hand with botulinum toxin injections. J Neurol Neurosurg Psychiatry 52:355–363
242. Das TK, Park DM (1989) Botulinum toxin in treating spasticity. Br J Clin Pract 43:401–403
243. Carruthers JD, Carruthers JA (1992) Treatment of glabellar frown lines with C. botulinum A exotoxin. J Dermatol Surg Oncol 18:17–21
244. Mezaki T, Kaji R, Hamano T, Nagamine T, Shibasaki H, Shimizu T et al (1994) Optimisation of botulinum treatment for cervical and axial dystonias: experience with a Japanese type A toxin. J Neurol Neurosurg Psychiatry 57:1535–1537
245. Greene P, Kang U, Fahn S, Brin M, Moskowitz C, Flaster E (1990) Double-blind, placebo-controlled trial of botulinum toxin injections for the treatment of spasmodic torticollis. Neurology 40:1213–1218
246. Lew MF, Adornato BT, Duane DD, Dykstra DD, Factor SA, Massey JM et al (1997) Botulinum toxin type B: a double-blind, placebo-controlled, safety and efficacy study in cervical dystonia. Neurology 49:701–707
247. Brashear A, Lew MF, Dykstra DD, Comella CL, Factor SA, Rodnitzky RL et al (1999) Safety and efficacy of NeuroBloc (botulinum toxin type B) in type A-responsive cervical dystonia. Neurology 53:1439–1446
248. Truong D, Duane DD, Jankovic J, Singer C, Seeberger LC, Comella CL et al (2005) Efficacy and safety of botulinum type A toxin (Dysport) in cervical dystonia: results of the first US randomized, double-blind, placebo-controlled study. Mov Disord 20:783–791
249. Simpson DM, Alexander DN, O'Brien CF, Tagliati M, Aswad AS, Leon JM et al (1996) Botulinum toxin type A in the treatment of upper extremity spasticity: a randomized, double-blind, placebo-controlled trial. Neurology 46:1306–1310
250. Hesse S, Reiter F, Konrad M, Jahnke MT (1998) Botulinum toxin type A and short-term electrical stimulation in the treatment of upper limb flexor spasticity after stroke: a randomized, double-blind, placebo-controlled trial. Clin Rehabil 12:381–388

251. Smith SJ, Ellis E, White S, Moore AP (2000) A double-blind placebo-controlled study of botulinum toxin in upper limb spasticity after stroke or head injury. Clin Rehabil 14:5–13
252. Bhakta BB, Cozens JA, Chamberlain MA, Bamford JM (2000) Impact of botulinum toxin type A on disability and carer burden due to arm spasticity after stroke: a randomised double blind placebo controlled trial. J Neurol Neurosurg Psychiatry 69:217–221
253. Bakheit AM, Thilmann AF, Ward AB, Poewe W, Wissel J, Muller J et al (2000) A randomized, double-blind, placebo-controlled, dose-ranging study to compare the efficacy and safety of three doses of botulinum toxin type A (Dysport) with placebo in upper limb spasticity after stroke. Stroke 31:2402–2406
254. Bakheit AM, Pittock S, Moore AP, Wurker M, Otto S, Erbguth F et al (2001) A randomized, double-blind, placebo-controlled study of the efficacy and safety of botulinum toxin type A in upper limb spasticity in patients with stroke. Eur J Neurol 8:559–565
255. Childers MK, Brashear A, Jozefczyk P, Reding M, Alexander D, Good D et al (2004) Dose-dependent response to intramuscular botulinum toxin type A for upper-limb spasticity in patients after a stroke. Arch Phys Med Rehabil 85:1063–1069
256. Suputtitada A, Suwanwela NC (2005) The lowest effective dose of botulinum A toxin in adult patients with upper limb spasticity. Disabil Rehabil 27:176–184
257. Richardson D, Sheean G, Werring D, Desai M, Edwards S, Greenwood R et al (2000) Evaluating the role of botulinum toxin in the management of focal hypertonia in adults. J Neurol Neurosurg Psychiatry 69:499–506
258. Pittock SJ, Moore AP, Hardiman O, Ehler E, Kovac M, Bojakowski J et al (2003) A double-blind randomised placebo-controlled evaluation of three doses of botulinum toxin type A (Dysport) in the treatment of spastic equinovarus deformity after stroke. Cerebrovasc Dis 15:289–300
259. Sutherland DH, Kaufman KR, Wyatt MP, Chambers HG, Mubarak SJ (1999) Double-blind study of botulinum A toxin injections into the gastrocnemius muscle in patients with cerebral palsy. Gait Posture 10:1–9
260. Ubhi T, Bhakta BB, Ives HL, Allgar V, Roussounis SH (2000) Randomised double blind placebo controlled trial of the effect of botulinum toxin on walking in cerebral palsy. Arch Dis Child 83:481–487
261. Baker R, Jasinski M, Maciag-Tymecka I, Michalowska-Mrozek J, Bonikowski M, Carr L et al (2002) Botulinum toxin treatment of spasticity in diplegic cerebral palsy: a randomized, double-blind, placebo-controlled, dose-ranging study. Dev Med Child Neurol 44:666–675
262. Barwood S, Baillieu C, Boyd R, Brereton K, Low J, Nattrass G et al (2000) Analgesic effects of botulinum toxin A: a randomized, placebo-controlled clinical trial. Dev Med Child Neurol 42:116–121
263. Mall V, Heinen F, Siebel A, Bertram C, Hafkemeyer U, Wissel J et al (2006) Treatment of adductor spasticity with BTX-A in children with CP: a randomized, double-blind, placebo-controlled study. Dev Med Child Neurol 48:10–13
264. Bjornson K, Hays R, Graubert C, Price R, Won F, McLaughlin JF et al (2007) Botulinum toxin for spasticity in children with cerebral palsy: a comprehensive evaluation. Pediatrics 120:49–58
265. Kawamura A, Campbell K, Lam-Damji S, Fehlings D (2007) A randomized controlled trial comparing botulinum toxin A dosage in the upper extremity of children with spasticity. Dev Med Child Neurol 49:331–337
266. Moore AP, de-Hall RA, Smith CT, Rosenbloom L, Walsh HP, Mohamed K et al (2008) Two-year placebo-controlled trial of botulinum toxin A for leg spasticity in cerebral palsy. Neurology 71:122–128

Chapter 4
Clinical Use of Botulinum Neurotoxin: Autonomic Conditions

Dirk Dressler

Abstract Botulinum neurotoxins inhibit the release from cholinergic nerve terminals of the sympathetic and parasympathetic autonomic nervous systems. This has clinical utility in treating conditions involving hyperactivity of the autonomic nervous system, including hyperhidrosis, hypersalivation and conditions of smooth muscle hyperactivity. These clinical uses of the neurotoxin are reviewed in this chapter.

Keywords Botulinum neurotoxin · Autonomic · Hyperhidrosis · Hypersalivation · Sympathetic · Parasympathetic · Detrusor

4.1 Anatomy

The autonomic nervous system, also called the visceral or vegetative nervous system, innervates all inner organs via a dense network of slow conducting nerve fibres. Its function is the—mostly involuntary—maintenance of the equilibrium of body functions under changing environmental conditions. In general, the sympathetic part of the autonomic nervous system adapts the organism to 'fight or flight', whereas the parasympathetic part adapts it to 'rest and digest'. The particular effects of the autonomic nervous system upon the effector organs are shown in Table 4.1.

The autonomic nervous system can be divided into a central part and a peripheral part. Its central part is not well understood. Its main components are the nuclei tractus solitarii, the formatio reticularis and the hypothalamus from where it connects to the hypophysis and other parts of the brain. Its peripheral part consists of afferent fibres called viscerosensory fibres mainly travelling with the sympathetic nerves and entering the spinal cord via the posterior roots. Their cell body is located within the spinal ganglions. Efferent fibres originate from the spinal cord and can be divided into sympathetic and parasympathetic ones. The recent discovery of nitric oxide as a transmitter suggests expanding this concept to include a third efferent pathway. The efferent peripheral autonomic pathways, which transmit virtually all efferences

D. Dressler (✉)
Movement Disorders Section, Department of Neurology, Hannover Medical School,
Carl-Neuberg-Str. 1, 30625 Hannover, Germany
e-mail: dressler.dirk@mh-hannover.de

K. A. Foster (ed.), *Clinical Applications of Botulinum Neurotoxin,*
Current Topics in Neurotoxicity 5, DOI 10.1007/978-1-4939-0261-3_4,

Table 4.1 Effects of the autonomic nervous system upon its effector organs

Effector organ	Parasympathetic nervous system	Sympathetic nervous system
Eye	Pupil constriction Accommodation increases	Pupil dilatation Accommodation decreases
Lacrimal glands	Tear production increases	Tear production decreases
Salivary glands	Saliva production increases	Saliva production decreases
Sweat glands		Sweat production increases
Arteries/skin		Constriction Perfusion decreases 'Centralisation' of blood flow
Arteries/intestinal		Constriction Perfusion decreases Blood pressure increases
Arteries/muscular		Perfusion increases
Arteries/kidney		Perfusion decreases
Heart	Heart rate decreases Cardiac output decreases	Heart rate increases Contractility increases Cardiac output increases
Lung	Breathing rate decreases Bronchial constriction	Breathing rate increases Bronchial dilatation
Stomach	Motility increases Acid production increases	Motility decreases
Intestine	Motility increases	Motility decreases
Liver	Glycogenolysis Gluconeogenesis	
Bladder	Detrusor increases Sphincter decreases	Detrusor decreases Sphincter increases
Genitals	Sexual arousal/erection	Orgasm/ejaculation
Adrenal gland		Noradrenaline/epinephrine secretion

except the innervation of striatal muscles, originate from the spinal cord and the brainstem. As shown in Fig. 4.1, the first peripheral autonomic neuron is always cholinergic.

In the sympathetic nervous system, the first peripheral neuron originates from the thoracic and lumbar spinal cord ('thoracolumbar origin'). The second one is located in the paravertebral truncus sympathicus, consisting of its prevertebral ganglia, the ganglion cervicalis superior and the ganglion stellatum and the prevertrebral ganglia consisting of the ganglion coelicaum, the ganglion mesentericum superius and the ganglion mesentericum inferius. The second peripheral neuron is adrenergic and reaches the effector organ. Only the innervation of the sweat glands is cholinergic, thus allowing therapeutic modulation by botulinum toxin (BT). The adrenal medulla is directly innervated by cholinergic first peripheral neurons.

In the parasympathetic nervous system, the first peripheral neuron originates from the brainstem (nucleus Edinger–Westphal, nuclei salivatorii, nucleus dorsalis nervi vagi) or the sacral spinal cord ('craniosacral origin'). The second peripheral neuron is also cholinergic. It is located close to the effector organ. For the pupil, it is located in the ganglion ciliary, for the glandula parotis, in the ganglion oticum and for the glandulae submandibularis and sublinguales, in the ganglion submandibularis. For the heart, lung, stomach, liver, pancreas, kidney and intestine, all innervated by the vagal nerve, and for the rectum, bladder and genitals, the second neuron is located within these organs.

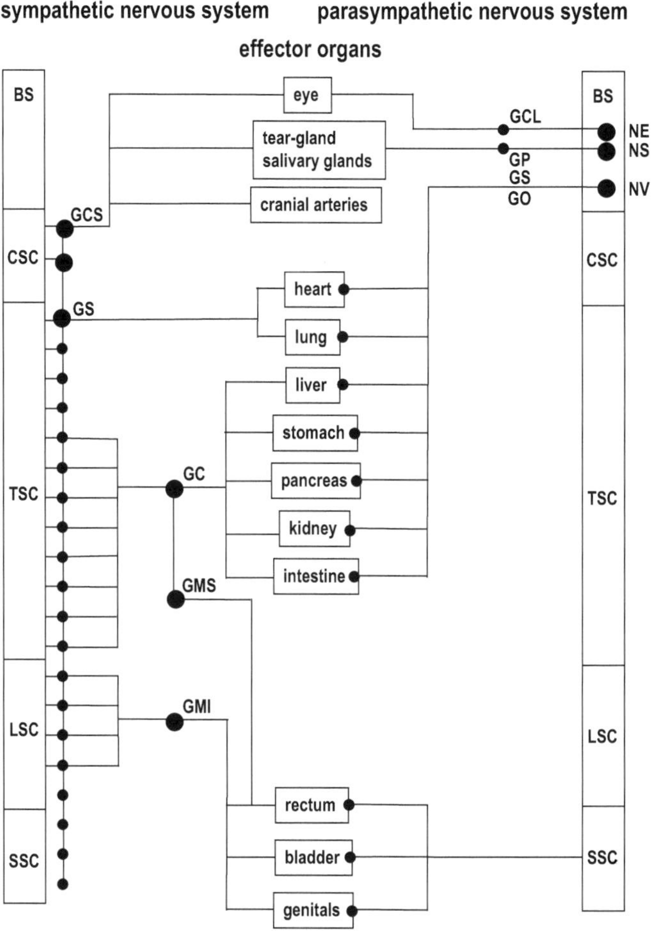

Fig. 4.1 Autonomic nervous system: overview. *BS* brainstem, *CSC* cervical spinal cord, *GC* ganglion coeliacum, *GCL* ganglion ciliary, *GCS* ganglion cervicalis superior, *GMI* ganglion mesentericum inferius, *GMS* ganglion mesentericum superius, *GO* ganglion oticum, *GP* ganglion paroticum, *GS* ganglion stellatum, *LSC* lumbar spinal cord, *NE* Nucleus Edinger–Westphal, *NS* nuclei salivatorii, *NV* nucleus dorsalis nervi vagi, *SSC* sacral spinal cord, *TSC* thoracic spinal cord. (Modified after: Kahle W, Leonhardt H, Platzer W (1979) Taschenatlas der Anatomie für Studium und Praxis. Band 3: Nervensystem und Sinnesorgane. 3 überarbeitete Auflage. Thieme-Verlag, Stuttgart)

4.2   Bladder Dysfunctions (see also Chap. 5)

4.2.1   Anatomy and Physiology of the Lower Urinary Tract

The lower urinary tract consists of the bladder and the urethra. The bladder is emptied by activation of the muscle fibres of the bladder wall, also called M. detrusor vesicae, and by relaxation of the bladder sphincters, i.e. the internal sphincter, formed by muscle fibres of the bladder wall and supported by the M. puboversicalis, and the external sphincter, formed by muscle fibres of the pelvic floor and supported by the M. sphincter urethrae.

The M. detrusor vesicae, the internal sphincter and the M. sphincter urethrae are controlled by the autonomic nervous system. Its sympathetic part activates the internal sphincter and the M. sphincter urethrae and inhibits the M. detrusor vesicae putting the bladder into 'storage mode'. Its parasympathetic part activates the M. detrusor vesicae and relaxes the internal sphincter and the M. sphincter urethrae putting the bladder into 'micturition mode'. The parasympathetic fibres of the lower urinary tract originate in the sacral spinal cord, their second peripheral neuron in ganglia located close to the bladder. The sympathetic fibres originate in the lumbar spinal cord, their second peripheral neuron in the ganglion mesentericum inferior. The external sphincter consists of striated muscle fibres and is under direct control of the frontal lobe micturition centre via the pyramidal tract and the nucleus Onuf in the sacral spinal cord. The central control of the autonomic innervation of the lower urinary tract is performed by the pontine micturition centre coordinating the sympathetic and parasympathetic efferences.

4.2.2 Detrusor Sphincter Dyssynergia

Detrusor sphincter dyssynergia (DSD) is defined as an incoordinated action of the detrusor and the sphincter muscles of the bladder. It is caused by central nervous system dysfunction and leads to residual urine after micturition causing urinary tract infection, renal damage and urosepsis.

Therapeutic interventions attempt to release the residual urine by catherisation (intermittent or permanent) or reduction of the sphincter tonus by medication or surgery. Problems of these therapies include infections, inadequate efficacy and incontinence. BT therapy for DSD was introduced as early as 1988 by Dennis Dykstra and collaborators [67]. Subsequently, numerous studies [66], [219], [26], [80], [55], [81] confirm robust effects on urethral pressure, post-micturition residual urine volume and bladder pressure for approximately 60–90 days. BT is applied as 100–250 mouse units (MU) Botox® or 150 MU Dysport®. Studies using Neurobloc®/MyoBloc® or Xeomin® have not been published yet. BT injections are performed either transurethrally using cystoscopy or transperineally using electromyography. Other methods for approaching the target muscles include ultrasound [45], magnetic resonance imaging [220] and fluoroscopic techniques [242].

4.2.3 Idiopathic Detrusor Overactivity

Idiopathic detrusor overactivity, also called overactive bladder or urge syndrome, is defined as an urgency to micturate in the absence of pathological processes. Additional symptoms may include incontinence, pollakisuria and nocturia. Prevalence of idiopathic detrusor overactivity is considerably high and constantly increasing with age. Conventional therapy is based upon anticholinergic drugs. Although effective, this treatment option frequently produces cumbersome systemic anticholinergic adverse effects, especially in the elderly.

BT injections into the detrusor muscle produce robust therapeutic effects on urinary frequency, urgency, incontinence, quality of life and urodynamic parameters lasting for 3–9 months [144], [91], [206], [39], [138], [112], [204], [203], [133], [186]. Two studies are randomised controlled studies [82], [205]. Most studies used Botox®. Use of Dysport® or Myobloc/Neurobloc® was rare. BT doses range from 50 to 300 MU Botox® with most studies applying 200 MU. Dysport® is used in doses of 500 MU. BT is spread over about 30 injection sites. Dilutions are usually 100 MU Botox®/10.0 ml of normal saline. NeuroBloc/MyoBloc®, although used in comparable doses, seems to produce shorter therapeutic effects than BT type A products [106]. When flexible cystoscopy is used for BT application, intravesical local anaesthesia is sufficient. Rigid cystoscopy requires general anaesthesia. Dosing seems critical so that urinary retention requiring clean intermittent catherisation is frequent. Urinary tract infection caused by the procedure is another frequent adverse effect, whereas haematuria is rare [137].

4.2.4 Neurogenic Detrusor Overactivity

Neurogenic detrusor overactivity describes the urgency to micturate usually caused by spinal cord lesions, less frequently by supraspinal lesions. Conventional treatment options are identical to idiopathic detrusor overactivity.

BT therapy is widely published and produces robust therapeutic effects on incontinence, maximum cystomeric capacity, maximum detrusor pressure and quality of life [221], [123], [192], [133], [14], [93], [229], [101], [125], [183], [136], [222], [200], [120], [216], [121], [84]. Two studies are randomised control studies [222], [68]. Target muscle is the detrusor vesicae muscle. The procedure is the same as in idiopathic detrusor overactivity. BT doses vary between 100 and 400 MU Botox® and 500 and 1,000 MU Dysport®. The duration of benefit ranges from 3 to 12 months. Treatment results and adverse effects seem to be similar to those seen in idiopathic detrusor overactivity, only that higher BT doses seem to be necessary [101]. Neurogenic as well as idiopathic bladder overactivity can also be treated in children with Botox® doses of 10–12 MU/kg body weight (up to 360 MU; [119], [107], [216], [195], [215], [214]) or Dysport® doses of 20 MU/kg body weight (up to 400 MU; [3]).

4.2.5 Urinary Retention

Urinary retention may occur in patients with paretic M. detrusor vesicae or overactive urethral sphincters. Causes include cauda equina lesions and peripheral polyneuropathy. BT injections into the external sphincter can improve voiding [182], [132], but also may fail [76].

4.2.6 Bladder Pain Syndrome

Bladder pain syndrome or interstitial cystitis has been treated with BT, with therapeutic effects on pain, daytime micturition, night-time micturition and maximal cystometric capacity [228], [134], [85], [147], [86]. The mechanism of action remains unclear.

4.3 Pelvic Floor Disorders

4.3.1 Pelvic Floor Spasms

Pelvic floor spasms include a number of heterogeneous pain conditions of unknown aetiology which are otherwise difficult to treat. They may respond to some degree to BT. Trials have been reported on vaginism [36], vestibulodynia or coital pain [38], vulvodynia or dyspareunia [96], chronic perineal pain, dysmenorrhoea, dyspareunia, dyschezia, nonmenstrual pelvic pain [1], [111], [240], outlet obstruction constipation [153] and anismus [102], [117], [199]. One study followed a randomised controlled design [1]. Target muscles include the levator ani, obturatorius internus, puborectalis and pubococcygeus muscles. Typical doses per target muscle were in the order of 20–40 MU Botox®.

4.3.2 Anal Fissures

Anal fissures describe painful rhagades of the perianal tissue originally caused by excessive stretching of the anal mucosa and then maintained by inflammation pain-induced increase of the anal tone. Reduction of the anal tone, therefore, offers a therapeutic option. Conventionally, this can be achieved by topical application of isosorbide dinitrate, glyceryl trinitrate, calcium channel blockers or by lateral sphincterectomy. Medical treatment has success rates in the order of 60–80 %. Whilst nitrates frequently produce headaches, calcium channel blockers are better tolerated. Lateral sphincterectomy produces even better therapeutic outcomes and has a low recurrence rate, but bears the risk of permanent incontinence.

An injection of 20–40 MU Botox® or 50–100 MU Dysport® into the M. sphincter ani externus or M. sphincter ani internus can be a therapeutic alternative with success rates between medical and surgical treatment [241], [94], [118], [74]. Often, singular injections may allow the anal rhagades to heal. Subsequent injections may be applied if necessary. Anal fissures may reoccur. Recurrence rates are higher after BT therapy than after lateral sphincterectomy. BT therapy can be accompanied by mild and transient incontinence. BT costs compared to topical medical treatment are a substantial disadvantage. Pros and cons of the available treatment options suggest a stepwise approach starting with calcium antagonists, escalating to BT therapy and eventually initiating surgery [253].

4.4 Prostate Disorders

4.4.1 Benign Prostate Hyperplasia

Benign prostate hyperplasia (BPH), less exactly also called benign prostate hypertrophy, describes proliferation of prostate connective tissue (stromal cells) and epithelial cells in the periurethral prostate (static component) as well as increased prostatic smooth muscle tone (dynamic component). BPH affects a large percentage of the ageing male population and leads to urethral obstruction with urinary retention, pollakisuria, dysuria, urolithiasis and increased risk of urinary tract infection [187]. BPH is believed to be associated with increased local testosterone levels. Conventional therapy includes alpha receptor blockers and anticholinergics for relaxation of intraprostatic smooth muscles as well as 5-alpha-reductase inhibitors to reduce testosterone production. Whilst medication is only partially effective and may produce adverse effects, removal or destruction of periurethral prostate tissue using various minimally invasive or surgical techniques is usually effective, but bears risks of incontinence and retrograde ejaculation.

BT can relax intraprostatic smooth muscles as well as reduce glandular secretion. Animal experiments also suggest induction of glandular apoptosis [56], [48]. Additionally, BT may improve urinary retention by reducing urethral sphincter tone.

BT, transurethrally usually given in doses of 100–200 MU Botox®, can significantly improve flow rate, prostate size and quality of life in BPH patients [154], [135], [46], [47], [48], [226].

4.5 Gastrointestinal Disorders

4.5.1 Gastroparesis

Gastroparesis describes a delayed gastric emptying of non-obstructive origin leading to postprandial nausea, bloating and early satiety. It is caused by dysfunction of the local autonomic nervous system and may be induced by diabetes mellitus, scleroderma and surgical procedures [258]. In about 30 % of cases, the underlying process is idiopathic [126]. Antiemetics including metoclopramide and domperidone as well as erythromycin may be helpful, but may also be accompanied by adverse effects. Invasive procedures include insertion of jejunostomic feeding tubes, partial gastrectomy and implantation of gastric stimulators.

An injection of 100–200 MU Botox® into the pyloric sphincter can improve cardinal complaints [73], [140], [158], [98], [110], [254], [141], [37], [27], [239], [64], [11], [124], [190], [88]. Two studies followed a randomised controlled design [12], [78].

4.5.2 Sphincter Oddi Spasms

Sphincter Oddi spasms are diagnosed when the sphincteric pressure rises to more than 40 mmHg [251]. They can produce pancreatitis and liver dysfunction. Endoscopic sphincterotomy [251] is the treatment of choice, but bears the risk of perforation and enterocholedochal reflux.

After an initial study performed as early as 1994 [176], several studies showed that 100 MU Botox® can improve the sequelae of sphincter Oddi spasms [249], [251]. Since BT therapy has to be repeated over a prolonged period of time, the use of BT injections as a diagnostic tool to prove the indication of sphincterotomy was suggested [77].

4.6 Oesophageal Disorders

4.6.1 Achalasia

Achalasia describes aperistalsis and reduced relaxation of the lower oesophageal sphincter (LES). Clinically, it manifests with progressive dysphagia to solids and liquids, retention of food and saliva, regurgitation, thoracic pain and weight loss. Its cause is unknown. Inflammation of the myenteric plexus with ganglial cell loss and fibrosis indicates sympathetic degeneration [90].

There is no causal treatment. Symptomatic treatment targets reduction of LES pressure and includes the laparoscopic Heller myotomy and endoscopic pneumatic dilatations. They are successful in approximately 80 % of the cases [197]. Newer approaches include endoscopic myotomy and self-expanding metal stents. The most reliable treatment for achalasia is myotomy followed by dilatation. Both treatments are well tolerated [99]. Pharmacotherapy, including calcium channel blocking agents, isosorbide dinitrate, nitroglycerine, anticholinergics and beta-adrenergic agonists, does not produce satisfactory results [7].

BT was introduced as a treatment for achalasia by Pasricha and his group in 1994 [177]. An injection of 50–200 MU Botox® (usually 80–100 MU) into all four LES quadrants can improve achalasia for 3–9 months [177], [178], [198], [8], [75],

[179], [52], [92], [9], [129], [162], [185], [245], [250], [10], [5], [58], [83], [171], [235], [155], [257], [20]. Adverse effects are mild and transient and include thoracic pain and reflux. Two studies have been performed using 240–250 MU Dysport® for the treatment of achalsia [5], [156]. Results are similar to those using Botox®. BT seems to be reserved for patients who cannot undergo conventional therapies, including elderly patients, patients with comorbidity and patients with oesophageal perforation or epiphrenic diverticula [143]. BT therapy may, however, be combined with conventional therapies for achalasia [259], [15].

4.6.2 Cricopharyngeal Achalasia

In cricopharyngeal achalasia, the upper oesophageal sphincter (UES) is affected, either primary or secondary to various neurological conditions, including stroke and Parkinson's disease, to laryngectomy or to local tumours. UES achalasia can also be treated with myotomy and dilatation. Overall success rates are similar to LES achalasia [95].

UES achalsia can be treated successfully with 10–100 MU Botox® [65], [51], [13], [31], [35], [4], [2], [224], [100], [175], [159], [18] or with 30–120 MU Dysport® [211], [193], [194] applied to the cricopharyngeal muscle. The success rate is similar to myectomy or dilatation and tends to be reduced when complex pharyngo-oesophageal movement disorders are present [95].

4.6.3 Unspecific Oesophageal Spasms

BT injections into the LES have also been used successfully to treat rare forms of unspecific oesophageal spasms [157], [234].

4.7 Hyperhidrosis

4.7.1 Axillary Hyperhidrosis

Hyperhidrosis describes excessive sweating in the axillary region. It is almost entirely idiopathic, often with positive family history, juvenile onset and female preponderance at least in specialised hyperhidrosis clinics. Hyperhidrosis of the palms and soles may be associated. Additional involvement of other typical areas of sweating including the chest, the back or the head is rare. Axillary hyperhidrosis is, with 90 % of cases, by far the most common form of hyperhidrosis. Sweating is physiological and can be separated into thermoregular sweating, predominantly activating eccrine sweat glands, and emotional sweating, predominantly activating apocrine sweat glands. It is difficult to separate hyperhidrosis from normal sweating by abstract or quantitative definition. In clinical practice, however, sweating in hyperhidrotic patients is usually so strong that reference to quantitative definitions is unnecessary. The frequency of hyperhidrosis is reported to be 2.8 % in the general US population [236]. Hyperhidrosis is medically benign, but may be socially devastating.

Conventional therapies include topical antiperspirants such as aluminium chloride, iontophoresis, anticholinergic drugs and surgery such as retrodermal axillary curettage and endoscopic thoracic sympathectomy (ETS) for palmar and axillary hyperhidrosis. Benzodiazepines, clonidine and non-steroidal anti-inflammatory drugs may have supportive antihydrotic effects. Topical antiperspirants and iontophoresis have short-term effectivity only, and skin irritation may occur. Anticholinergics are usually mildly effective only and frequently produce severe systemic adverse effects. ETS requires a major operation associated with intraoperative risks of bleeding, pneumothorax and haematothorax and post-operative risks of chest pain and compensatory hyperhidrosis elsewhere in the body.

Axillary hyperhidrosis can be effectively treated with multiple intradermal or subdermal injections of BT typically placed about 2 cm apart from each other in the hyperhidrotic skin area. For this, total doses per axilla of 50–100 MU Botox® [40], [166], [104], [167], [149], [150], [61], 100–200 MU Dysport® [212], [213], [207], [105], [160], 2,000–5,000 MU MyoBloc®/Neurobloc® [59], [170] or 100 MU Xeomin® [61] may be used. In almost all patients, hyperhidrosis can be abolished. Adverse effects are virtually nil, except for Neurobloc®/MyoBloc® which may produce autonomic adverse effects [59], [60]. Skin lesions due to dryness of skin do not occur. Injection site pain is unpleasant, but tolerable without further treatment. Due to its disadvantageous pH value, Neurobloc®/MyoBloc® produces increased injection site pain [59]. The duration of the therapeutic effect is often reported to be longer than the 12 weeks typically seen in motor indications. Not surprisingly, a recent formalised assessment of the American Academy of Neurology confirmed that BT therapy is a safe and effective therapy of axillary hyperhidrosis [168].

Future research needs to address the prolonged duration of action of BT for treatment of hyperhidrosis. It should also systematically study dose and dilution optimization for all available products.

4.7.2 *Palmar Hyperhidrosis*

BT may be successfully used for treatment of palmar hyperhidrosis. For this, 30–160 MU Botox® per palm may be used [169], [163], [243], [201], [181], [231], [225], [148], [227], [256]. Usually, BT doses applied per palm were 50 or 100 MU Botox®, 120–280 MU Dysport® [213], [227], 100 MU Xeomin® [61] or 4,000–5,000 MU Neurobloc®/MyoBloc® [23], [21], [60]. With this, palmar hyperhidrosis can be reduced substantially. Sometimes, for anatomical reasons, hyperhidrotic areas remain.

The major problem of BT treatment of palmar hyperhidrosis is injection pain in the sensitive fingertips. Several approaches have been suggested to reduce this pain, including cryoanalgesia [230], [22], [196], [145], ulnar and median nerve blocks [41], topical analgesics [30, 180], intravenous regional anaesthesia [33] and iontophoretic BT application [53]. Our experience indicates that ischaemic blockade induced by a proximal arm cuff produces sufficient anaesthesia together with prevention of haemorrhagic BT wash-out. Speedy BT application of less than 45 s per palm also helps improve compliance. Occasionally, mild transient hand paresis may occur. Compensatory sweating elsewhere in the body does not occur.

4.7.3 Plantar Hyperhidrosis

Plantar hyperhidrosis can be safely and effectively treated with 50–100 MU Botox$^®$ per planta [165], [244], [223], [42]. Injection pain is a major unsolved problem.

4.7.4 Diffuse Sweating

Diffuse sweating affects the head, the chest and the back. It may also affect, to a lesser degree, the axillae, the palms and the feet. Often, it is symptomatic, caused by infections (viral, bacterial, especially tuberculosis and malaria), endocrine dysfunction (hyperthyroidism, hyperpituitarism, diabetes mellitus, menopause and pregnancy, phaeochromocytoma, carcinoid syndrome, acromegaly), neurological disorders (parkinsonism), malignancies (myeloproliferative syndromes, Hodgkin's disease), collagenosis, drugs (antidepressants, Acyclovir, Ciprofloxacin, etc.), intoxication and withdrawal (alcohol, heroin, cocaine and other substances). Treatment is predominantly causal, if possible. Symptomatic conventional therapy is similar to axillary hyperhidrosis, only more problematic due to its more widespread distribution.

Frontal hyperhidrosis can be treated with 20–90 MU Botox$^®$ [127], [208] and cranial hyperhidrosis with around 300 MU Xeomin$^®$ [61]. Caution has to be applied to prevent paretic effects upon mimic muscles. Other hyperhidrotic skin areas can also be treated with BT in the same way [[24], Dressler unpublished observations].

4.7.5 Frey's Syndrome

Frey's syndrome, also named gustatory sweating or auriculotemporal syndrome, describes sweating, flushing and erythema of the temporal skin in patients who underwent parotidectomy. After parotidectomy, the parasympathetic nerve fibres originally innervating the parotid gland may aberrantly sprout into the sweat glands and vessels of the temporal skin. Eating, physiologically activating the parotid gland, then induces sweating and flushing. Depending on the detection method, Frey's syndrome may affect almost all patients with parotidectomy. It is, by far, the most common sequelae of parotidectomy. Although medically benign, it may profoundly affect the patient's social interactions. Conventional treatment includes various surgical interventions, radiation, anticholinergics and topical antiperspirants. It is either of limited efficacy or accompanied by problematic adverse effects.

BT injected into the affected skin produces a safe and reliable relief. Depending upon the affected skin area, 2.5–150 MU Botox® [218], [54], [29], [164], [142], [25], [139], [247], [63], around 2,500 MU Myobloc®/NeurBloc® [43] and 60–80 MU Dysport® [218] may be applied.

4.8 Hypersalivation

Hypersalivation describes the presence of excessive saliva in the mouth which may cause the patient to drool and result in severe embarrassment. Almost always, hypersalivation is caused by impaired swallowing of saliva as in parkinsonian syndromes, in motor neuron disease (amyotrophic lateral sclerosis) and cerebral palsy. Rarely, it may be caused by a genuine hyperproduction of saliva as sometimes seen with the administration of neuroleptic drugs. Conventional therapy includes muscarinic anticholinergic drugs, such as atropine, scopolamine, tricyclic antidepressants for reduction of watery secretion, beta receptor blocking agent for reduction of mucous secretion, radiotherapy or resection of the parotid and submandibular glands, ligation of their glandular ducts, neurectomy of the tympanic nerve, mucolytics and behavioural therapy.

BT therapy of hypersalivation targets the paired parotid glands, producing approximately 30 % of the saliva, and the paired submandibular glands, producing approximately 70 % of the saliva. The paired sublingual glands are difficult to target. With a production of less than 5 % of the saliva, they are not used for BT therapy. BT injections into the parotid gland are easy to place when anatomical landmarks are used. BT placement into the sublingual glands is also easy to perform, although ultrasound guidance has been suggested [114]. The minor salivary glands are spread over the oral cavity and cannot be targeted with BT.

4.8.1 Hypersalivation in Cerebral Palsy

BT has been applied successfully for drooling in children with cerebral palsy, usually under general anaesthesia. Several studies have shown the efficacy of Botox® [113], [71], [237], [115], [116], [209], [16], [191], [252], [6], [173], [210], [232], [255] and Myobloc®/NeuroBloc® [252], [189], [19]. Recommended BT doses depend on the child's body weight. For Botox®, they are 10–50 MU (parotid gland) and 10–50 MU (submandibular gland), for Dysport®, 15–75 MU (parotid gland) and 15–75

MU (submandibular gland) and for Myobloc®/NeuroBloc®, 400–1,000 MU (parotid gland) and 250–1,000 MU (submandibular gland; [188]). Botox® doses of 5–10 MU per parotid seem ineffective [34], [103]. Adverse effects are rare and may consist of dysphagia, weakness of jaw closure, increased saliva viscosity, excessive dryness of mouth and parotitis. Adjunct therapies may include speech therapy, occupational therapy, physiotherapy and behavioural therapy.

4.8.2 Hypersalivation in Parkinsonian Syndromes

BT therapy can be used successfully to reduce hypersalivation in patients with parkinsonian syndromes. For this, a total dose of 10–80 MU Botox® (usually 50–80 MU) is injected into both parotid glands [174], [79], [44], [57]. If additional submandibular injections are performed, the total doses are between 50 and 100 MU with 2/3 of the total dose injected into the parotid gland and 1/3 into the submandibular gland [72], [184]. When Dysport® is used, the total doses injected into both parotid glands are 20–300 MU [28], [89], [146]. When the submandibular glands are to be injected additionally, the total dose is 450 MU [151].

4.8.3 Hypersalivation in Motor Neuron Disease (Amyothrophic Lateral Sclerosis)

BT can be successfully used to treat hypersalivation in motor neuron disease, although disorders of the motor neuron are usually considered contraindications. BT is placed into both parotid glands, sometimes additionally in both submandibular glands. Adverse effects are mild and transient and similar to those seen in patients with parkinsonian syndromes. They included mild chewing difficulties, mild dysphagia and viscous saliva.

For this, Botox® is used in total doses of 12–140 MU with the majority placed in the parotid gland [87], [184], [152], [246]. Usual doses are around 30–40 MU for each parotid gland and around 10 MU for each submandibular gland. Myobloc®/NeuroBloc® is used in total doses of 2,500 MU with 1,000 MU applied to both parotid glands and 2,500 MU to both submandibular glands [109], [49], [50]. Dysport® is used in total doses of 40–150 MU injected into both parotid glands [146].

4.8.4 Hypersalivation Due to Administration of Neuroleptic Drugs

For treatment of hypersalivation induced by the atypical neuroleptic drug clozapine, Myobloc®/NeuroBloc® is injected successfully in both parotid glands in total doses of 2,000 MU and into both submandibular glands in total doses of 500 MU [233].

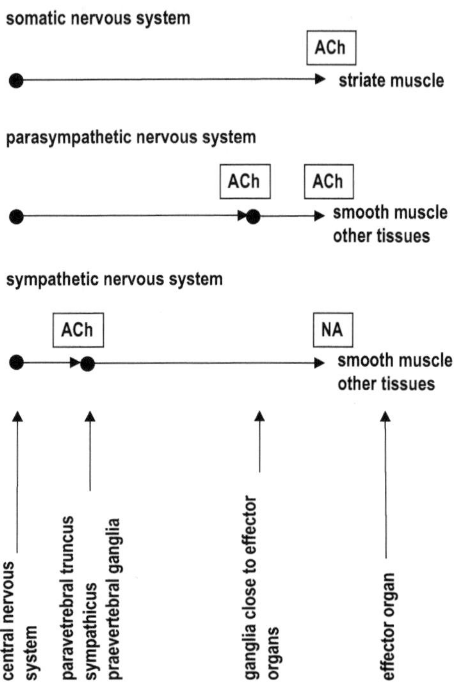

Fig. 4.2 Neurotransmitters in the peripheral somatic and autonomic nervous system. *ACh* acetylcholine, *NA* noradrenalin

4.8.5   *Hypersalivation in Various Ear–Nose–Throat Conditions*

Occasionally, BT may be used to treat hypersalivation caused by carcinomas of the larynx, pharynx, parotid glands and connective tissues of the neck [69], [70], [202].

4.9   Hyperlacrimation

4.9.1   *Crocodile Tears Syndrome*

Crocodile tears syndrome describes the uncontrolled flow of tears during eating in patients with facial nerve impairment. It is caused by aberrant sprouting of autonomic facial nerve fibres originally innervating salivary glands. The condition is rare and medically benign. Usually, it is caused by Bell's palsy or traumatic facial palsy; rarely, it maybe congenital. It is named after the observation that crocodiles produce tears when chewing compresses their tear glands [248].

Crocodile tears syndrome can be effectively treated with BT injections directly into the lacrimal gland. For this, 2.5 MU Botox® [108], [17], [128], [248] or 20 MU Dysport® [32], [122], [161] has been used. Occasional ptosis seems to be the only adverse effect. Lacrimal gland injections may also be used for hyperlacrimation after submandibular gland autografts and entropion.

4.10   Other Conditions

4.10.1   Reynaud Phenomenon

BT has been used with controversial results in patients with Raynaud phenomenon [238], [130], [172].

References

1. Abbott JA, Jarvis SK, Lyons SD, Thomson A, Vancaille TG (2006) Botulinum toxin type A for chronic pain and pelvic floor spasm in women: a randomized controlled trial. Obstet Gynecol 108:915–923
2. Ahsan SF, Meleca RJ, Dworkin JP (2000) Botulinum toxin injection of the cricopharyngeus muscle for the treatment of dysphagia. Otolaryngol Head Neck Surg 122:691–695
3. Akbar M, Abel R, Seyler TM, Bedke J, Haferkamp A, Gerner HJ, Möhring K (2007) Repeated botulinum-A toxin injections in the treatment of myelodysplastic children and patients with spinal cord injuries with neurogenic bladder dysfunction. BJU Int 100:639–645
4. Alberty J, Oelerich M, Ludwig K, Hartmann S, Stoll W (2000) Efficacy of botulinum toxin A for treatment of upper esophageal sphincter dysfunction. Laryngoscope 110:1151–1156
5. Allescher HD, Storr M, Seige M, Gonzales-Donoso R, Ott R, Born P, Frimberger E, Weigert N, Stier A, Kurjak M, Rösch T, Classen M (2001) Treatment of achalasia: botulinum toxin injection vs pneumatic balloon dilation. A prospective study with long-term follow-up. Endoscopy 33:1007–1017
6. Alrefai AH, Aburahma SK, Khader YS (2009) Treatment of sialorrhea in children with cerebral palsy: a double-blind placebo controlled trial. Clin Neurol Neurosurg 111:79–82
7. Annese V, Janssens J, Vantrappen G (1990) Primary esophageal motility disorders: medical treatment. Minerva Dietol Gastroenterol 36:145–155
8. Annese V, Basciani M, Perri F, Lombardi G, Frusciante V, Simone P, Andriulli A, Vantrappen G (1996) Controlled trial of botulinum toxin injection versus placebo and pneumatic dilation in achalasia. Gastroenterology 111:1418–1124
9. Annese V, Basciani M, Borrelli O, Leandro G, Simone P, Andriulli A (1998) Intrasphincteric injection of botulinum toxin is effective in long-term treatment of esophageal achalasia. Muscle Nerve 21:1540–1542
10. Annese V, Bassotti G, Coccia G, Dinelli M, D'Onofrio V, Gatto G, Leandro G, Repici A, Testoni PA, Andriulli A (2000) A multicentre randomised study of intrasphincteric botulinum toxin in patients with esophageal achalasia. Gut 46:597–600
11. Arts J, van Gool S, Caenepeel P, Verbeke K, Janssens J, Tack J (2006) Influence of intrapyloric botulinum toxin injection on gastric emptying and meal-related symptoms in gastroparesis patients. Aliment Pharmacol Ther 24:661–667
12. Arts J, Holvoet L, Caenepeel P, Bisschops R, Sifrim D, Verbeke K, Janssens J, Tack J (2007) Clinical trial: a randomized-controlled crossover study of intrapyloric injection of botulinum toxin in gastroparesis. Aliment Pharmacol Ther 26:1251–1258
13. Atkinson SI, Rees J (1997) Botulinum toxin for cricopharyngeal dysphagia: case reports of CT-guided injection. J Otolaryngol 26:273–276
14. Bagi P, Biering-Sørensen F (2004) Botulinum toxin A for treatment of neurogenic detrusor overactivity and incontinence in patients with spinal cord lesions. Scand J Urol Nephrol 38:495–498
15. Bakhshipour A, Rabbani R, Shirani S, Soleimani HA, Mikaeli J (2010) Comparison of pneumatic dilation with pneumatic dilation plus botulinum toxin for treatment of achalasia. Acta Med Iran 48:107–110

16. Banerjee KJ, Glasson C, O'Flaherty SJ (2006) Parotid and submandibular botulinum toxin A injections for sialorrhoea in children with cerebral palsy. Dev Med Child Neurol 48:883–887
17. Barañano DE, Miller NR (2004) Long term efficacy and safety of botulinum toxin A injection for crocodile tears syndrome. Br J Ophthalmol 88:588–589
18. Barnes MA, Ho AS, Malhotra PS, Koltai PJ, Messner AH (2011) The use of botulinum toxin for pediatric cricopharyngeal achalasia. Int J Pediatr Otorhinolaryngol 75:1210–1214
19. Basciani M, Di Rienzo F, Fontana A, Copetti M, Pellegrini F, Intiso D (2011) Botulinum toxin type B for sialorrhoea in children with cerebral palsy: a randomized trial comparing three doses. Dev Med Child Neurol 53:559–564
20. Bassotti G, D'Onofrio V, Battaglia E, Fiorella S, Dughera L, Iaquinto G, Mazzocchi A, Morelli A, Annese V (2006) Treatment with botulinum toxin of octo-nonagerians with oesophageal achalasia: a two-year follow-up study. Aliment Pharmacol Ther 23:1615–1619
21. Baumann LS, Halem ML (2003) Systemic adverse effects after botulinum toxin type B (myobloc) injections for the treatment of palmar hyperhidrosis. Arch Dermatol 139:226–227
22. Baumann L, Frankel S, Welsh E, Halem M (2003) Cryoanalgesia with dichlorotetrafluoroethane lessens the pain of botulinum toxin injections for the treatment of palmar hyperhidrosis. Dermatol Surg 29:1057–1059
23. Baumann L, Slezinger A, Halem M, Vujevich J, Mallin K, Charles C, Martin LK, Black L, Bryde J (2005) Double-blind, randomized, placebo-controlled pilot study of the safety and efficacy of Myobloc (botulinum toxin type B) for the treatment of palmar hyperhidrosis. Dermatol Surg 31:263–270
24. Bechara FG, Sand M, Achenbach RK, Sand D, Altmeyer P, Hoffmann K (2007) Focal hyperhidrosis of the anal fold: successful treatment with botulinum toxin A. Dermatol Surg 33:924–927
25. Beerens AJ, Snow GB (2002) Botulinum toxin A in the treatment of patients with Frey syndrome. Br J Surg 89:116–119
26. Beleggia F, Beccia E, Imbriani E, Basciani M, Intiso D, Cioffi R, Simone P, Ricci Barbini V (1997) The use of type A botulin toxin in the treatment of detrusor-sphincter dyssynergia. Arch Ital Urol Androl 69 (Suppl 1):61–63
27. Ben-Youssef R, Baron PW, Franco E, Walter MH, Lewis T, Ojogho O (2006) Intrapyloric injection of botulinum toxin a for the treatment of persistent gastroparesis following successful pancreas transplantation. Am J Transplant 6:214–218
28. Bhatia KP, Münchau A, Brown P (1999) Botulinum toxin is a useful treatment in excessive drooling in saliva. J Neurol Neurosurg Psychiatry 67:697
29. Bjerkhoel A, Trobbe O (1997) Frey's syndrome: treatment with botulinum toxin. J Laryngol Otol 111:839–844
30. Blaheta HJ, Vollert B, Zuder D, Rassner G (2002) Intravenous regional anesthesia (Bier's block) for botulinum toxin therapy of palmar hyperhidrosis is safe and effective. Dermatol Surg 28:666–671
31. Blitzer A, Brin MF (1997) Use of botulinum toxin for diagnosis and management of cricopharyngeal achalasia. Otolaryngol Head Neck Surg 116:328–330
32. Boroojerdi B, Ferbert A, Schwarz M, Herath H, Noth J (1998) Botulinum toxin treatment of synkinesia and hyperlacrimation after facial palsy. J Neurol Neurosurg Psychiatry 65:111–114
33. Bosdotter Enroth S, Rystedt A, Covaciu L, Hymnelius K, Rystedt E, Nyberg R, Naver H, Swartling C (2010) Bilateral forearm intravenous regional anesthesia with prilocaine for botulinum toxin treatment of palmar hyperhidrosis. J Am Acad Dermatol 63:466–474
34. Bothwell JE, Clarke K, Dooley JM et al (2002) Botulinum toxin A as a treatment for excessive drooling in children. Pediatr Neurol 27:18–22
35. Brant CQ, Siqueira ES, Ferrari AP Jr (1999) Botulinum toxin for oropharyngeal dysphagia: case report of flexible endoscope-guided injection. Dis Esophagus 12:68–73
36. Brin MF, Vapnek JM (1997) Treatment of vaginismus with botulinum toxin injections. Lancet 349:252–253
37. Bromer MQ, Friedenberg F, Miller LS, Fisher RS, Swartz K, Parkman HP (2005) Endoscopic pyloric injection of botulinum toxin A for the treatment of refractory gastroparesis. Gastrointest Endosc 61:833–839

38. Brown CS, Glazer HI, Vogt V, Menkes D, Bachmann G (2006) Subjective and objective outcomes of botulinum toxin type A treatment in vestibulodynia: pilot data. J Reprod Med 51:635–641
39. Brubaker L, Richter HE, Visco A, Mahajan S, Nygaard I, Braun TM, Barber MD, Menefee S, Schaffer J, Weber AM, Wei J (2008) Pelvic Floor Disorders Network Refractory idiopathic urge urinary incontinence and botulinum A injection. J Urol 180:217–222
40. Bushara KO, Park DM, Jones JC, Schutta HS (1996) Botulinum toxin-a possible new treatment for axillary hyperhidrosis. Clin Exp Dermatol 21:276–278
41. Campanati A, Lagalla G, Penna L, Gesuita R, Offidani A (2004) Local neural block at the wrist for treatment of palmar hyperhidrosis with botulinum toxin: technical improvements. J Am Acad Dermatol 51:345–348
42. Campanati A, Bernardini ML, Gesuita R, Offidani A (2007) Plantar focal idiopathic hyperhidrosis and botulinum toxin: a pilot study. Eur J Dermatol 17:52–54
43. Cantarella G, Berlusconi A, Mele V, Cogiamanian F, Barbieri S m (2010) Treatment of Frey's syndrome with botulinum toxin type B. Otolaryngol Head Neck Surg 143:214–218
44. Carod Artal FJ (2003) Treatment of sialorrhoea in neurological diseases with trans-dermic injections of botulinum toxin type A in the parotid glands. Neurologia 18:280–284
45. Chen SL, Bih LI, Chen GD, Huang YH, You YH, Lew HL (2010) Transrectal ultrasound-guided transperineal botulinum toxin a injection to the external urethral sphincter for treatment of detrusor external sphincter dyssynergia in patients with spinal cord injury. Arch Phys Med Rehabil 91:340–344
46. Chuang YC, Chiang PH, Huang CC, Yoshimura N, Chancellor MB (2005) Botulinum toxin type A improves benign prostatic hyperplasia symptoms in patients with small prostates. Urology 66:775–779
47. Chuang YC, Chiang PH, Yoshimura N, De Miguel F, Chancellor MB (2006a) Sustained beneficial effects of intraprostatic botulinum toxin type A on lower urinary tract symptoms and quality of life in men with benign prostatic hyperplasia. BJU Int 98:1033–1037
48. Chuang YC, Tu CH, Huang CC, Lin HJ, Chiang PH, Yoshimura N, Chancellor MB (2006b) Intraprostatic injection of botulinum toxin type-A relieves bladder outlet obstruction in human and induces prostate apoptosis in dogs. BMC Urol 6:12
49. Contarino MF, Pompili M, Tittoto P, Vanacore N, Sabatelli M, Cedrone A, Rapaccini GL, Gasbarrini G, Tonali PA, Bentivoglio AR (2007) Botulinum toxin B ultrasound-guided injections for sialorrhea in amyotrophic lateral sclerosis and Parkinson's disease. Parkinsonism Relat Disord 13:299–303
50. Costa J, Rocha ML, Ferreira J, Evangelista T, Coelho M, de Carvalho M (2008) Botulinum toxin type-B improves sialorrhea and quality of life in bulbaronset amyotrophic lateral sclerosis. J Neurol 255:545–550
51. Crary MA, Glowasky AL (1996) Using botulinum toxin A to improve speech and swallowing function following total laryngectomy. Arch Otolaryngol Head Neck Surg 122:760–763
52. Cuillière C, Ducrotté P, Zerbib F, Metmann EH, de Looze D, Guillemot F, Hudziak H, Lamouliatte H, Grimaud JC, Ropert A, Dapoigny M, Bost R, Lémann M, Bigard MA, Denis P, Auget JL, Galmiche JP, Bruley des Varannes S (1997) Achalasia: outcome of patients treated with intrasphincteric injection of botulinum toxin. Gut 41:87–92
53. Davarian S, Kalantari KK, Rezasoltani A, Rahimi A (2008) Effect and persistency of botulinum toxin iontophoresis in the treatment of palmar hyperhidrosis. Australas J Dermatol 49:75–79
54. de Bree R, Duyndam JE, Kuik DJ, Leemans CR (2009) Repeated botulinum toxin type A injections to treat patients with Frey syndrome. Arch Otolaryngol Head Neck Surg 135: 287–290
55. de Sèze M, Petit H, Gallien P, de Sèze MP, Joseph PA, Mazaux JM, Barat M (2002) Botulinum a toxin and detrusor sphincter dyssynergia: a double-blind lidocaine-controlled study in 13 patients with spinal cord disease. Eur Urol 42:56–62
56. Doggweiler R, Zermann DH, Ishigooka M, Schmidt RA (1998) Botox-induced prostatic involution. Prostate 37:44–50

57. Dogu O, Apaydin D, Sevim S, Talas DU, Aral M (2004) Ultrasound-guided versus 'blind' intraparotid injections of botulinum toxin-A for the treatment of sialorrhoea in patients with Parkinson's disease. Clin Neurol Neurosurg 106:93–96

58. D'Onofrio V, Miletto P, Leandro G, Iaquinto G (2001) Long-term follow-up of achalasia patients treated with botulinum toxin. Digest Liver Dis 33:105–110

59. Dressler D, Adib Saberi F, Benecke R (2002) Botulinum toxin type B for treatment of axillar hyperhidrosis. J Neurol 249:1729–1732

60. Dressler D, Benecke R (2003) Autonomic side effects of botulinum toxin type B treatment of cervical dystonia and hyperhidrosis. Eur Neurol 49:34–38

61. Dressler D (2010) Comparing Botox and Xeomin for axillar hyperhidrosis. J Neural Transm 117:317–319

62. Dressler D (2010) Routine use of Xeomin in patients previously treated with Botox: long term results. Eur J Neurol 16 (Suppl 2):2–5

63. Dulguerov P, Quinodoz D, Cosendal G, Piletta P, Lehmann W (2000). Frey syndrome treatment with botulinum toxin. Otolaryngol Head Neck Surg 122:821–827

64. Dumonceau JM, Giostra E, Bech C, Spahr L, Schroft A, Shah D (2006) Acute delayed gastric emptying after ablation of atrial fibrillation: treatment with botulinum toxin injection. Endoscopy 38:543

65. Dunne J, Hayes M, Cameron D (1993) Botulinum toxin A for cricopharyngeal dystonia. Lancet 342:559

66. Dykstra DD, Sidi AA (1990) Treatment of detrusor-sphincter dyssynergia with botulinum A toxin: a double-blind study. Arch Phys Med Rehabil 71:24–26

67. Dykstra DD, Sidi AA, Scott AB, Pagel JM, Goldish GD (1988) Effects of botulinum A toxin on detrusor-sphincter dyssynergia in spinal cord injury patients. J Urol 139:919–922

68. Ehren I, Volz D, Farrelly E, Berglund L, Brundin L, Hultling C, Lafolie P (2007) Efficacy and impact of botulinum toxin A on quality of life in patients with neurogenic detrusor overactivity: a randomised, placebo-controlled, double-blind study. Scand J Urol Nephrol 41:335–340

69. Ellies M, Laskawi R, Rohrbach-Volland S, Rödel R, Beuche W (2001) Blocking secretion of exocrine glands in the head-neck area by administration of botulinum toxin A. Therapy of a rare disease picture. HNO 49:807–813

70. Ellies M, Laskawi R, Rohrbach-Volland S, Arglebe C, Beuche W (2002) Botulinum toxin to reduce saliva flow: selected indications for ultrasound-guided toxin application into salivary glands. Laryngoscope 112:82–86

71. Ellies M, Rohrbach-Volland S, Arglebe C, Wilken B, Laskawi R, Hanefeld (2002) Successful management of drooling with botulinum toxin A in neurologically disabled children. Neuropediatrics 33:327–330

72. Ellies M, Laskawi R, Rohrbach-Volland S, Arglebe C (2003) Up-to-date report of botulinum toxin therapy in patients with drooling caused by different etiologies. J Oral Maxillofac Surg 61:454–457

73. Ezzeddine D, Jit R, Katz N, Gopalswamy N, Bhutani MS (2002) Pyloric injection of botulinum toxin for treatment of diabetic gastroparesis. Gastrointest Endosc 55:920–923

74. Festen S, Gisbertz SS, van Schaagen F, Gerhards MF (2009) Blinded randomized clinical trial of botulinum toxin versus isosorbide dinitrate ointment for treatment of anal fissure. Br J Surg 96:1393–1399

75. Fishman VM, Parkman HP, Schiano TD, Hills C, Dabezies MA, Cohen S, Fisher RS, Miller LS (1996) Symptomatic improvement in achalasia after botulinum toxin injection of the lower esophageal sphincter. Am J Gastroenterol 91:1724–1730

76. Fowler CJ, Betts CD, Christmas TJ, Swash M, Fowler CG (1992) Botulinum toxin in the treatment of chronic urinary retention in women. Br J Urol 70:387–389

77. Friedenberg F, Gollamudi S, Parkman HP (2004) The use of botulinum toxin for the treatment of gastrointestinal motility disorders. Dig Dis Sci 49:165–175

78. Friedenberg FK, Palit A, Parkman HP, Hanlon A, Nelson DB (2008) Botulinum toxin A for the treatment of delayed gastric emptying. Am J Gastroenterol 103:416–423

79. Friedman A, Potulska A (2001) Botulinum toxin for treatment of parkinsonian sialorrhea. Neurol Neurochir Pol 35(Suppl 3):23–27
80. Gallien P, Robineau S, Verin M, Le Bot MP, Nicolas B, Brissot R (1998) Treatment of detrusor sphincter dyssynergia by transperineal injection of botulinum toxin. Arch Phys Med Rehabil 79:715–717
81. Gallien P, Reymann JM, Amarenco G, Nicolas B, de Sèze M, Bellissant E (2005) Placebo controlled, randomised, double blind study of the effects of botulinum A toxin on detrusor sphincter dyssynergia in multiple sclerosis patients. J Neurol Neurosurg Psychiatry 76: 1670–1676
82. Ghei M, Maraj BH, Miller R, Nathan S, O'Sullivan C, Fowler CJ, Shah PJ, Malone-Lee J (2005) Effects of botulinum toxin B on refractory detrusor overactivity: a randomized, double-blind, placebo controlled, crossover trial. Urol 174:1873–1877
83. Ghoshal UC, Chaudhuri S, Pal BB, Dhar K, Ray G, Banerjee PK (2001) Randomized controlled trial of intrasphincteric botulinum toxin A injection versus balloon dilatation in treatment of achalasia. Dis Esophagus 14:227–231
84. Giannantoni A, Mearini E, Di Stasi SM, Costantini E, Zucchi A, Mearini L, Fornetti P, Del Zingaro M, Navarra P, Porena M (2004) New therapeutic options for refractory neurogenic detrusor overactivity. Minerva Urol Nefrol 56:79–87
85. Giannantoni A, Costantini E, Di Stasi SM, Tascini MC, Bini V, Porena M (2006) Botulinum A toxin intravesical injections in the treatment of painful bladder syndrome: a pilot study. Eur Urol 49:704–709
86. Giannantoni A, Porena M, Costantini E, Zucchi A, Mearini L, Mearini E (2008) Botulinum A toxin intravesical injection in patients with painful bladder syndrome: 1-year followup. J Urol 179:1031–1034
87. Giess R, Naumann M, Werner E, Riemann R, Beck M, Puls I, Reiners C, Toyka KV (2000) Injections of botulinum toxin A into the salivary glands improve sialorrhoea in amyotrophic lateral sclerosis. J Neurol Neurosurg Psychiatry 69:121–123
88. Gil RA, Hwynn N, Fabian T, Joseph S, Fernandez HH (2011) Botulinum toxin type A for the treatment of gastroparesis in Parkinson's disease patients. Parkinsonism Relat Disord 17:285–287
89. Glickman S, Deaney CN (2001) Treatment of relative sialorrhoea with botulinum toxin type A: description and rationale for an injection procedure with case report. Eur J Neurol 8:567–571
90. Goldblum JR, Rice TW, Richter JE (1996) Histopathologic features in esophagomyotomy specimens from patients with achalasia. Gastroenterology 111:648–654
91. Gomez CS, Kanagarajah P, Gousse A (2010) The use of botulinum toxin a in idiopathic overactive bladder syndrome. Curr Urol Rep 11:353–359
92. Gordon JM, Eaker EY (1997) Prospective study of esophageal botulinum toxin injection in high risk achalasia patients. Am J Gastroenterol 92:1812–1817
93. Grosse J, Kramer G, Stöhrer M (2005) Success of repeat detrusor injections of botulinum A toxin in patients with severe neurogenic detrusor overactivity and incontinence. Eur Urol 47:653–659
94. Gui D, Cassetta E, Anastasio G, Bentivoglio AR, Maria G, Albanese A (1994). Botulinum toxin for chronic anal fissure. Lancet 344(8930):1127–1128
95. Gui D, Rossi S, Runfola M, Magalini SC (2003) Review article: botulinum toxin in the therapy of gastrointestinal motility disorders. Aliment Pharmacol Ther 18:1–16
96. Gunter J, Brewer A, Tawfik O (2004) Botulinum toxin A for vulvodynia: a case report. J Pain 5:238–240
97. Guntinas-Lichius O, Eckel HE (2002) Temporary reduction of salivation in laryngectomy patients with pharyngocutaneous fistulas by botulinum toxin A injection. Laryngoscope 112:187–189
98. Gupta P, Rao SS (2002) Attenuation of isolated pyloric pressure waves in gastroparesis in response to botulinum toxin injection: a case report. Gastrointest Endosc 56:770–772
99. Gutschow CA, Töx U, Leers J, Schäfer H, Prenzel KL, Hölscher AH (2010) Botox, dilation, or myotomy? Clinical outcome of interventional and surgical therapies for achalasia. Langenbecks Arch Surg 395:1093–1099

100. Haapaniemi JJ, Laurikainen EA, Pulkkinen J, Marttila RJ (2001) Botulinum toxin in the treatment of cricopharyngeal dysphagia. Dysphagia 16:171–175
101. Hajebrahimi S, Altaweel W, Cadoret J, Cohen E, Corcos J (2005) Efficacy of botulinum-A toxin in adults with neurogenic overactive bladder: initial results. Can J Urol 12:2543–2546
102. Hallan RI, Williams NS, Melling J, Waldron DJ, Womack NR, Morrison JF (1988) Treatment of anismus in intractable constipation with botulinum A toxin. Lancet 8613:714–717
103. Hassin-Baer S, Scheuer E, Buchman AS, Jacobson I, Ben-Zeev B (2005) Botulinum toxin injections for children with excessive drooling. J Child Neurol 20:120–123
104. Heckmann M, Ceballos-Baumann AO, Plewig G (2001) Botulinum toxin A for axillary hyperhidrosis (excessive sweating). N Engl J Med 344:48893
105. Heckmann M, Plewig G, Hyperhidrosis Study Group (2005) Low-dose efficacy of botulinum toxin A for axillary hyperhidrosis: a randomized, side-by-side, open-label study. Arch Dermatol 141:1255–1259
106. Hirst GR, Watkins AJ, Guerrero K, Wareham K, Emery SJ, Jones DR, Lucas MG (2007) Botulinum toxin B is not an effective treatment of refractory overactive bladder. Urology 69:69–73
107. Hoebeke P, De Caestecker K, Vande Walle J, Dehoorne J, Raes A, Verleyen P, Van Laecke E (2006) The effect of botulinum-A toxin in incontinent children with therapy resistant overactive detrusor. J Urol 176:328–330
108. Hofmann RJ (2000) Treatment of Frey's syndrome (gustatory sweating) and 'crocodile tears' (gustatory epiphora) with purified botulinum toxin. Ophthal Plast Reconstr Surg 16:289–291
109. Jackson CE, Gronseth G, Rosenfeld J, Barohn RJ, Dubinsky R, Simpson CB, McVey A, Kittrell PP, King R, Herbelin L, Muscle Study Group (2009) Randomized double-blind study of botulinum toxin type B for sialorrhea in ALS patients. Muscle Nerve 39:137–143
110. James AN, Ryan JP, Parkman HP (2003) Inhibitory effects of botulinum toxin on pyloric and antral smooth muscle. Am J Physiol Gastrointest Liver Physiol 285:G291–297
111. Jarvis SK, Abbott JA, Lenart MB, Steensma A, Vancaillie TG (2004) Pilot study of botulinum toxin type A in the treatment of chronic pelvic pain associated with spasm of the levator ani muscles. Aust N Z J Obstet Gynaecol 44:46–50
112. Jeffery S, Fynes M, Lee F, Wang K, Williams L, Morley R (2007) Efficacy and complications of intradetrusor injection with botulinum toxin A in patients with refractory idiopathic detrusor overactivity. BJU Int 100:1302–136
113. Jongerius PH, Rotteveel JJ, van den Hoogen F, Joosten F, van Hulst K, Gabreëls FJ (2001) Botulinum toxin A: a new option for treatment of drooling in children with cerebral palsy. Presentation of a case series. Eur J Pediatr 160:509–512
114. Jongerius PH, Joosten F, Hoogen FJ, Gabreels FJ, Rotteveel JJ (2003) The treatment of drooling by ultrasound-guided intraglandular injections of botulinum toxin type A into the salivary glands. Laryngoscope 113:107–111
115. Jongerius PH, Rotteveel JJ, van Limbeek J, Gabreëls FJ, van Hulst K, van den Hoogen FJ (2004a) Botulinum toxin effect on salivary flow rate in children with cerebral palsy. Neurology 63:1371–1375
116. Jongerius PH, van den Hoogen FJ, van Limbeek J, Gabreëls FJ, van Hulst K, Rotteveel JJ (2004b) Effect of botulinum toxin in the treatment of drooling: a controlled clinical trial. Pediatrics 114:620–627
117. Joo JS, Agachan F, Wolff B, Nogueras JJ, Wexner SD (1996) Initial North American experience with botulinum toxin type A for treatment of anismus. Dis Colon Rectum 39:1107–1111
118. Jost WH, Schimrigk K (1994) Therapy of anal fissure using botulin toxin. Dis Colon Drectum 37:1321–1324
119. Kajbafzadeh AM, Moosavi S, Tajik P, Arshadi H, Payabvash S, Salmasi AH, Akbari HR, Nejat F (2006) Intravesical injection of botulinum toxin type A: management of neuropathic bladder and bowel dysfunction in children with myelomeningocele. Urology 68:1091–1096
120. Kalsi V, Apostolidis A, Popat R, Gonzales G, Fowler CJ, Dasgupta P (2006) Quality of life changes in patients with neurogenic versus idiopathic detrusor overactivity after intradetrusor

injections of botulinum neurotoxin type A and correlations with lower urinary tract symptoms and urodynamic changes. Eur Urol 49:528–535

121. Karsenty G, Reitz A, Lindemann G, Boy S, Schurch B (2006) Persistence of therapeutic effect after repeated injections of botulinum toxin type A to treat incontinence due to neurogenic detrusor overactivity. Urology 68:1193–1197

122. Keegan DJ, Geerling G, Lee JP, Blake G, Collin JR, Plant GT (2002) Botulinum toxin treatment for hyperlacrimation secondary to aberrant regenerated seventh nerve palsy or salivary gland transplantation. Br J Ophthalmol 86:43–46

123. Kennelly MJ, Kang J (2003) Botulinum-A toxin injections as a treatment fro refractory detrusor hyperreflexia. Top Spinal Cord Inj Rehabil 8:46–53

124. Kent MS, Pennathur A, Fabian T, McKelvey A, Schuchert MJ, Luketich JD, Landreneau RJ (2007) A pilot study of botulinum toxin injection for the treatment of delayed gastric emptying following esophagectomy. Surg Endosc 21:754–757

125. Kessler TM, Danuser H, Schumacher M, Studer UE, Burkhard FC (2005) Botulinum A toxin injections into the detrusor: an effective treatment in idiopathic and neurogenic detrusor overactivity? Neurourol Urodyn 24:231–236

126. Khoo J, Rayner CK, Jones KL, Horowitz M (2009) Pathophysiology and management of gastroparesis. Expert Rev Gastroenterol Hepatol 3:167–181

127. Kinkelin I, Hund M, Naumann M, Hamm H (2000) Effective treatment of frontal hyperhidrosis with botulinum toxin A. Br J Dermatol 143:824–827

128. Kizkin S, Doganay S, Ozisik HI, Ozcan C (2005) Crocodile tears syndrome: botulinum toxin treatment under EMG guidance. Funct Neurol 20:35–37

129. Kolbasnik J, Waterfall WE, Fachnie B, Chen Y, Tougas G (1999) Long-term efficacy of botulinum toxin in classical achalasia: a prospective study. Am J Gastroenterol 94:3434–3439

130. Kossintseva I, Barankin B (2008) Improvement in both Raynaud disease and hyperhidrosis in response to botulinum toxin type A treatment. J Cutan Med Surg 12:189–193

131. Kroupa R, Hep A, Dolina J, Valek V, Matyasova Z, Prokesova J, Mrazova J, Sedmik J, Novotny I (2010) Combined treatment of achalasia-botulinum toxin injection followed by pneumatic dilatation: long-term results. Dis Esophagus 23:100–105

132. Kuo HC (2003) Effect of botulinum a toxin in the treatment of voiding dysfunction due to detrusor underactivity. Urology 61:550–554

133. Kuo HC (2004) Urodynamic evidence of effectiveness of botulinum A toxin injection in treatment of detrusor overactivity refractory to anticholinergic agents. Urology 63:868–872

134. Kuo HC (2005a) Preliminary results of suburothelial injection of botulinum a toxin in the treatment of chronic interstitial cystitis. Urol Int 75:170–174

135. Kuo HC (2005b) Prostate botulinum A toxin injection-an alternative treatment for benign prostatic obstruction in poor surgical candidates. Urology 65:670–674

136. Kuo HC (2006) Therapeutic effects of suburothelial injection of botulinum a toxin for neurogenic detrusor overactivity due to chronic cerebrovascular accident and spinal cord lesions. Urology 67:232–236

137. Kuo HC, Liao CH, Chung SD (2010) Adverse events of intravesical botulinum toxin a injections for idiopathic detrusor overactivity: risk factors and influence on treatment outcome. Eur Urol 8:919–926

138. Kuschel S, Werner M, Schmid DM, Faust E, Schuessler B (2008) Botulinum toxin-A for idiopathic overactivity of the vesical detrusor: a 2-year follow-up. Int Urogynecol J Pelvic Floor Dysfunct 19:905–909

139. Kyrmizakis DE, Pangalos A, Papadakis CE, Logothetis J, Maroudias NJ, Helidonis ES (2004) The use of botulinum toxin type A in the treatment of Frey and crocodile tears syndromes. J Oral Maxillofac Surg 62:840–844

140. Lacy BE, Zayat EN, Crowell MD, Schuster MM (2002) Botulinum toxin for the treatment of gastroparesis: a preliminary report. Am J Gastroenterol 97:1548–1552

141. Lacy BE, Crowell MD, Schettler-Duncan A, Mathis C, Pasricha PJ (2004) The treatment of diabetic gastroparesis with botulinum toxin injection of the pylorus. Diabetes Care 27: 2341–2347

142. Laskawi R, Drobik C, Schönebeck C (1998) Up-to-date report of botulinum toxin type A treatment in patients with gustatory sweating (Frey's syndrome). Laryngoscope 108:381–384

143. Leyden JE, Moss AC, MacMathuna P (2006) Endoscopic pneumatic dilation versus botulinum toxin injection in the management of primary achalasia. Cochrane Database Syst Rev 18:CD005046

144. Lie KY, Wong MY, Ng LG (2010) Botulinum toxin a for idiopathic detrusor overactivity. Ann Acad Med Singapore 39:714–715

145. Lim EC, Seet RC (2007) Another injection-free method to effect analgesia when injecting botulinum toxin for palmar hyperhidrosis: cryoanalgesia. Dermatol Online J 13:25

146. Lipp A, Trottenberg T, Schink T, Kupsch A, Arnold G (2003) A randomized trial of botulinum toxin A for treatment of drooling. Neurology 61:1279–1281

147. Liu HT, Kuo HC (2007) Intravesical botulinum toxin A injections plus hydrodistension can reduce nerve growth factor production and control bladder pain in interstitial cystitis. Urology 70:463–468

148. Lowe NJ, Yamauchi PS, Lask GP, Patnaik R, Iyer S (2002) Efficacy and safety of botulinum toxin type a in the treatment of palmar hyperhidrosis: a double-blind, randomized, placebo-controlled study. Dermatol Surg 28:822–827

149. Lowe PL, Cerdan-Sanz S, Lowe NJ (2003) Botulinum toxin type A in the treatment of bilateral primary axillary hyperhidrosis; efficacy and duration with repeated treatments. Dermatol Surg 29:5458

150. Lowe NJ, Glaser DA, Eadie N, Daggett S, Kowalski JW, Lai PY, North American Botox in Primary Axillary Hyperhidrosis Clinical Study Group (2007) Botulinum toxin type A in the treatment of primary axillary hyperhidrosis: a 52-week multicenter double-blind, randomized, placebo-controlled study of efficacy and safety. J Am Acad Dermatol 56:604–611

151. Mancini F, Zangaglia R, Cristina S, Sommaruga MG, Martignoni E, Nappi G, Pacchetti C (2003) Double-blind, placebo-controlled study to evaluate the efficacy and safety of botulinum toxin type A in the treatment of drooling in parkinsonism. Mov Disord 18:685–688

152. Manrique D (2005) Application of botulinum toxin to reduce the saliva in patients with amyotrophic lateral sclerosis. Rev Bras Otorrinolaringol 71:556–559

153. Maria G, Brisinda G, Bentivoglio AR, Cassetta E, Albanese A (2000) Botulinum toxin in the treatment of outlet obstruction constipation caused by puborectalis syndrome. Dis Colon Rectum 43:376–380

154. Maria G, Brisinda G, Civello IM, Bentivoglio AR, Sganga G, Albanese A (2003) Relief by botulinum toxin of voiding dysfunction due to benign prostatic hyperplasia: results of a randomized, placebo-controlled study. Urology 62:259–264

155. Martínek J, Siroký M, Plottová Z, Bures J, Hep A, Spicák J (2003) Treatment of patients with achalasia with botulinum toxin: a multicenter prospective cohort study. Dis Esophagus 16:204–209

156. Mikaeli J, Fazel A, Montazeri G, Yaghoobi M, Malekzadeh R (2001) Randomized controlled trial comparing botulinum toxin injection to pneumatic dilatation for the treatment of achalasia. Aliment Pharmacol Ther 15:1389–1396

157. Miller LS, Parkman HP, Schiano TD, Cassidy MJ, Ter RB, Dabezies MA, Cohen S, Fisher RS (1996) Treatment of symptomatic nonachalasia esophageal motor disorders with botulinum toxin injection at the lower esophageal sphincter. Dig Dis Sci 41:2025–2031

158. Miller LS, Szych GA, Kantor SB, Bromer MQ, Knight LC, Maurer AH, Fisher RS, Parkman HP (2002) Treatment of idiopathic gastroparesis with injection of botulinum toxin into the pyloric sphincter muscle. Am J Gastroenterol 97:1653–1660

159. Moerman M, Callier Y, Dick C, Vermeersch H (2002) Botulinum toxin for dysphagia due to cricopharyngeal dysfunction. Eur Arch Otorhinolaryngol 259:1–3

160. Moffat CE, Hayes WG, Nyamekye IK (2009) Durability of botulinum toxin treatment for axillary hyperhidrosis. Eur J Vasc Endovasc Surg 38:188–191

161. Montoya FJ, Riddell CE, Caesar R, Hague S (2002) Treatment of gustatory hyperlacrimation (crocodile tears) with injection of botulinum toxin into the lacrimal gland. Eye (Lond) 16: 705–709

162. Muehldorfer SM, Schneider TH, Hochberger J, Martus P, Hahn EG, Ell C (1999) Esophageal achalasia: intrasphincteric injection of botulinum toxin A versus balloon dilation. Endoscopy 31:517–521
163. Naumann M, Flachenecker P, Bröcker EB, Toyka KV, Reiners K (1997) Botulinum toxin for palmar hyperhidrosis. Lancet 349:252
164. Naumann M, Zellner M, Toyka KV, Reiners K (1997) Treatment of gustatory sweating with botulinum toxin. Ann Neurol 42:973–975
165. Naumann M, Hofmann U, Bergmann I, Hamm H, Toyka KV, Reiners K (1998) Focal hyperhidrosis: effective treatment with intracutaneous botulinum toxin. Arch Dermatol 134:301–304
166. Naumann M, Lowe NJ (on behalf of the BOTOX Hyperhidrosis Clinical Study Group) (2001) Botulinum toxin type A in treatment of bilateral primary axillary hyperhidrosis: randomised, parallel group, double blind, placebo controlled trial. BMJ 323:596–599
167. Naumann M, Lowe NJ, Kumar CR, Hamm H (on behalf of the BOTOX Hyperhidrosis Clinical Study Group) (2003) Botulinum toxin type A is a safe and effective treatment for axillary hyperhidrosis over 16 months: a prospective study. Arch Dermatol 139:731–736
168. Naumann M, So Y, Argoff CE, Childers MK, Dykstra DD, Gronseth GS, Jabbari B, Kaufmann HC, Schurch B, Silberstein SD, Simpson DM, Therapeutics and Technology Assessment Subcommittee of the American Academy of Neurology (2008) Assessment: Botulinum neurotoxin in the treatment of autonomic disorders and pain (an evidence-based review): report of the Therapeutics and Technology Assessment Subcommittee of the American Academy of Neurology. Neurology 70:1707–1714
169. Naver H, Swartling C, Aquilonius SM (2000) Palmar and axillary hyperhidrosis treated with botulinum toxin: one-year clinical follow-up. Eur J Neurol 7:55–62
170. Nelson L, Bachoo P, Holmes J (2005) Botulinum toxin type B: a new therapy for axillary hyperhidrosis. Br J Plast Surg 58:228–232
171. Neubrand M, Scheurlen C, Schepke M, Sauerbruch T (2002) Long-term results and prognostic factors in the treatment of achalasia with botulinum toxin. Endoscopy 34:519–523
172. Neumeister MW (2010) Botulinum toxin type A in the treatment of Raynaud's phenomenon. J Hand Surg Am 35:2085–2092
173. Ong LC, Wong SW, Hamid HA (2009) Treatment of drooling in children with cerebral palsy using ultrasound guided intraglandular injections of botulinum toxin A. J Pediatr Neurol 7:141–145
174. Pal PK, Calne DB, Calne S, Tsui JK (2000) Botulinum toxin A as treatment for drooling saliva in PD.Neurology 54:244–247
175. Parameswaran MS, Soliman AM (2002) Endoscopic botulinum toxin injection for cricopharyngeal dysphagia. Ann Otol Rhinol Laryngol 111:871–874
176. Pasricha PJ, Miskovsky EP, Kalloo AN (1994) Intrasphincteric injection of botulinum toxin for suspected sphincter of Oddi dysfunction. Gut 35:1319–1321
177. Pasricha PJ, Ravich WJ, Hendix RT, Sostre S, Jones B, Kalloo AN (1994) Treatment of achalasia with intrasphincteric injection of botulinum toxin: a pilot trial. Ann Intern Med 121:590–591
178. Pasricha PJ, Ravich WJ, Hendix RT, Sostre S, Jones B, Kalloo AN (1995) Intrasphincteric botulinum toxin for the treatment of achalasia. N Engl J Med 322:774–778
179. Pasricha PJ, Rai R, Ravich WJ, Hendix RT, Kalloo AN (1996) Botulinum toxin for achalasia: long-term outcome and predictors of response. Gastroenterology 110:1410–1415
180. Patel R, Halem M, Zaiac M (2009) The combined use of forced cold air and topical anesthetic cream for analgesia during the treatment of palmar hyperhydrosis with botulinum toxin injections. J Drugs Dermatol 8:948–951
181. Pérez-Bernal AM, Avalos-Peralta P, Moreno-Ramírez D, Camacho F (2005) Treatment of palmar hyperhidrosis with botulinum toxin type A: 44 months of experience. J Cosmet Dermatol 4:163–166
182. Phelan MW, Franks M, Somogyi GT, Yokoyama T, Fraser MO, Lavelle JP, Yoshimura N, Chancellor MB (2001) Botulinum toxin urethral sphincter injection to restore bladder emptying in men and women with voiding dysfunction. J Urol 165:1107–1110

183. Popat R, Apostolidis A, Kalsi V, Gonzales G, Fowler CJ, Dasgupta P (2005) A comparison between the response of patients with idiopathic detrusor overactivity and neurogenic detrusor overactivity to the first intradetrusor injection of botulinum-A toxin. J Urol 174:984–989

184. Porta M, Gamba M, Bertacchi G, Vaj P (2001) Treatment of sialorrhoea with ultrasound guided botulinum toxin type A injection in patients with neurological disorders. J Neurol Neurosurg Psychiatry 70:538–540

185. Prakash C, Freedland KE, Chan MF, Clouse RE (1999). Botulinum toxin injections for achalasia can approximate the short term efficacy of a single pneumatic dilation: a survival analysis approach. Am J Gastroenterol 94:328–333

186. Rajkumar GN, Small DR, Mustafa AW, Conn G (2005) A prospective study to evaluate the safety, tolerability, efficacy and durability of response of intravesical injection of botulinum toxin type A into detrusor muscle in patients with refractory idiopathic detrusor overactivity. BJU Int 96:848–852

187. Randall A (1931) Surgical pathology of prostatic obstruction. Williams and Wilkins, Baltimore

188. Reddihough D, Erasmus CE, Johnson H, McKellar GM, Jongerius PH (2010) Botulinum toxin assessment, intervention and aftercare for paediatric and adult drooling: international consensus statement. Eur J Neurol 17 (Suppl 2):109–121

189. Reddihough D, Graham HK (2011) Botulinum toxin type B for sialorrhea in children with cerebral palsy. Dev Med Child Neurol 53:488–489

190. Reddymasu SC, Singh S, Sankula R, Lavenbarg TA, Olyaee M, McCallum RW (2009) Endoscopic pyloric injection of botulinum toxin-A for the treatment of postvagotomy gastroparesis. Am J Med Sci 337:161–164

191. Reid SM, Johnstone BR, Westbury C, Rawicki B, Reddihough D (2008) Randomized trial of botulinum toxin injections into the salivary glands to reduce drooling in children with neurological disorders. Dev Med Child Neurol 50:123–128

192. Reitz A, Stöhrer M, Kramer G, Del Popolo G, Chartier-Kastler E, Pannek J, Burgdörfer H, Göcking K, Madersbacher H, Schumacher S, Richter R, von Tobel J, Schurch B (2004) European experience of 200 cases treated with botulinum-A toxin injections into the detrusor muscle for urinary incontinence due to neurogenic detrusor overactivity. Eur Urol 45:510–515

193. Restivo DA, Marchese Ragona R, Staffieri A, de Grandis D (2000) Successful botulinum toxin treatment of dysphagia in oculopharyngeal muscular dystrophy. Gastroenterology 119:1416

194. Restivo DA, Palmeri A, Marchese-Ragona R (2002) Botulinum toxin for cricopharyngeal dysfunction in Parkinson's disease. N Engl J Med 346:1174–1175

195. Riccabona M, Koen M, Schindler M, Goedele B, Pycha A, Lusuardi L, Bauer SB (2004) Botulinum-A toxin injection into the detrusor: a safe alternative in the treatment of children with myelomeningocele with detrusor hyperreflexia. J Urol 171:845–848

196. Richards RN (2009) Ethyl chloride spray for sensory relief for botulinum toxin injections of the hands and feet. J Cutan Med Surg 13:253–256

197. Richter JE (2010) Achalasia-an update. J Neurogastroenterol Motil 16:232–242

198. Rollan A, Gonzalez R, Carvajal S, Chianale J (1995) Endoscopic intrasphincteric injection of botulinum toxin for the treatment of achalasia. J Clin Gastroenterol 20:189–191

199. Ron Y, Avni Y, Lukovetski A, Wardi J, Geva D, Birkenfeld S, Halpern Z (2001) Botulinum toxin type-A in therapy of patients with anismus. Dis Colon Rectum 44:1821–1826

200. Ruffion A, Capelle O, Paparel P, Leriche B, Leriche A, Grise P (2006) What is the optimum dose of type A botulinum toxin for treating neurogenic bladder overactivity? BJU Int 97: 1030–1034

201. Saadia D, Voustianiouk A, Wang AK, Kaufmann H (2001) Botulinum toxin type A in primary palmar hyperhidrosis: randomized, single-blind, two-dose study. Neurology 57:2095–2099

202. Saberi FA, Schade H, Dressler D (2008) Botulinum toxin to treat parotid saliva retention after transient postoperative Stenon's duct occlusion. Toxicon 51 (Suppl 1):44

203. Sahai A (2006) A prospective study to evaluate the safety, tolerability, efficacy and durability of response of intravesical injection of botulinum toxin type A into detrusor muscle in patients with refractory idiopathic detrusor overactivity. BJU Int 97:413

204. Sahai A, Khan MS, Dasgupta P (2007) Efficacy of botulinum toxin-A for treating idiopathic detrusor overactivity: results from a single center, randomized, double-blind, placebo controlled trial. J Urol 177:2231–2236
205. Sahai A, Dowson C, Khan MS, Dasgupta P (2009) Improvement in quality of life after botulinum toxin-A injections for idiopathic detrusor overactivity: results from a randomized double-blind placebo-controlled trial. BJU Int 103:1509–1515
206. Sahai A, Dowson C, Khan MS, Dasgupta P, GKT Botulinum Study Group (2010) Repeated injections of botulinum toxin-A for idiopathic detrusor overactivity. Urology 75:552–558
207. Salmanpoor R, Rahmanian MJ (2002) Treatment of axillary hyperhidrosis with botulinum-A toxin. Int J Dermatol 41:428–430
208. Santana-Rodríguez N, Clavo-Varas B, Ponce-González MA, Jarabo-Sarceda JR, Pérez-Alonso D, Ruiz-Caballero JA, Olmo-Quintana V, Atallah Yordi N, Fiuza-Pérez MD (2010) Primary frontal hyperhidrosis successfully treated with low doses of botulinum toxin A as a useful alternative to surgical treatment. J Dermatolog Treat 23:49–51
209. Savarese R, Diamond M, Elovic E, Millis SR (2004) Intraparotid injection of botulinum toxin A as a treatment to control sialorrhea in children with cerebral palsy. Am J Phys Med Rehabil 83:304–311
210. Scheffer AR, Erasmus C, van Hulst K, van Limbeek J, Jongerius PH, van den Hoogen FJ (2010) Efficacy and duration of botulinum toxin treatment for drooling in 131 children. Arch Otolaryngol Head Neck Surg 136:873–877
211. Schneider I, Thumfart WF, Pototschnig C, Eckel HE (1994) Treatment of dysfunction of the cricopharyngeal muscle with botulinum A toxin: introduction of a new, noninvasive method. Ann Otol Rhinol Laryngol 103:31–35
212. Schnider P, Binder M, Kittler H, Birner P, Starkel D, Wolff K, Auff E (1999) A randomized, double-blind, placebo-controlled trial of botulinum A toxin for severe axillary hyperhidrosis. Br J Dermatol 140:677–680
213. Schnider P, Moraru E, Kittler H, Binder M, Kranz G, Voller B, Auff E (2001) Treatment of focal hyperhidrosis with botulinum toxin type A: long-term follow-up in 61 patients. Br J Dermatol 145:289–293
214. Schulte-Baukloh H, Michael T, Schobert J, Stolze T, Knispel HH (2002) Efficacy of botulinum-a toxin in children with detrusor hyperreflexia due to myelomeningocele: preliminary results. Urology 59:325–327
215. Schulte-Baukloh H, Michael T, Stürzebecher B, Knispel HH (2003) Botulinum-a toxin detrusor injection as a novel approach in the treatment of bladder spasticity in children with neurogenic bladder. Eur Urol 44:139–143
216. Schulte-Baukloh H, Knispel HH, Stolze T, Weiss C, Michael T, Miller K (2005) Repeated botulinum-A toxin injections in treatment of children with neurogenic detrusor overactivity. Urology 66:865–870
217. Schulte-Baukloh H, Schobert J, Stolze T, Stürzebecher B, Weiss C, Knispel HH (2005) Efficacy of botulinum-A toxin bladder injections for the treatment of neurogenic detrusor overactivity in multiple sclerosis patients: an objective and subjective analysis. Neurourol Urodyn 25:110–115
218. Schulze-Bonhage A, Schroder M, Ferbert A (1996) Botulinum toxin in the therapy of gustatory sweating. J Neurol 243:143–146
219. Schurch B, Hauri D, Rodic B, Curt A, Meyer M, Rossier AB (1996) Botulinum-A toxin as a treatment of detrusor-sphincter dyssynergia: a prospective study in 24 spinal cord injury patients. J Urol 155:1023–1029
220. Schurch B, Hodler J, Rodic B (1997) Botulinum A toxin as a treatment of detrusor-sphincter dyssynergia in patients with spinal cord injury: MRI controlled transperineal injections. J Neurol Neurosurg Psychiatry 63:474–476
221. Schurch B, Stöhrer M, Kramer G, Schmid DM, Gaul G, Hauri D (2000) Botulinum-A toxin for treating detrusor hyperreflexia in spinal cord injured patients: a new alternative to anticholinergic drugs? Preliminary results. J Urol 164:692–697

222. Schurch B, de Sèze M, Denys P, Chartier-Kastler E, Haab F, Everaert K, Plante P, Perrouin-Verbe B, Kumar C, Fraczek S, Brin MF, Botox Detrusor Hyperreflexia Study Team (2005) Botulinum toxin type a is a safe and effective treatment for neurogenic urinary incontinence: results of a single treatment, randomized, placebo controlled 6-month study. J Urol 174: 196–200
223. Sevim S, Dogu O, Kaleagasi H (2002) Botulinum toxin-A therapy for palmar and plantar hyperhidrosis. Acta Neurol Belg 102:167–170
224. Shaw GY, Searl JP (2001) Botulinum toxin treatment for cricopharyngeal dysfunction. Dysphagia 16:161–167
225. Shelley WB, Talanin NY, Shelley ED (1998) Botulinum toxin therapy for palmar hyperhidrosis.J Am Acad Dermatol. 38:227–229
226. Silva J, Silva C, Saraiva L, Silva A, Pinto R, Dinis P, Cruz F (2008) Intraprostatic botulinum toxin type a injection in patients unfit for surgery presenting with refractory urinary retention and benign prostatic enlargement. Effect on prostate volume and micturition resumption. Eur Urol 53:153–159
227. Simonetta Moreau M, Cauhepe C, Magues JP, Senard JM (2003) A double-blind, randomized, comparative study of Dysport vs. Botox in primary palmar hyperhidrosis. Br J Dermatol 149:1041–1045
228. Smith CP, Radziszewski P, Borkowski A, Somogyi GT, Boone TB, Chancellor MB (2004) Botulinum toxin A has antinociceptive effects in treating interstitial cystitis. Urology 64: 871–875
229. Smith CP, Nishiguchi J, O'Leary M, Yoshimura N, Chancellor MB (2005) Single-institution experience in 110 patients with botulinum toxin A injection into bladder or urethra. Urology 65:37–41
230. Smith KC, Comite SL, Storwick GS (2007) Ice minimizes discomfort associated with injection of botulinum toxin type A for the treatment of palmar and plantar hyperhidrosis. Dermatol Surg 33:S88–S91
231. Solomon BA, Hayman R (2000) Botulinum toxin type A therapy for palmar and digital hyperhidrosis. J Am Acad Dermatol 42:1026–1029
232. Sriskandan N, Moody A, Howlett DC (2010) Ultrasound-guided submandibular gland injection of botulinum toxin for hypersalivation in cerebral palsy. Br J Oral Maxillofac Surg 48:58–60
233. Steinlechner S, Klein C, Moser A, Lencer R, Hagenah J (2010) Botulinum toxin B as an effective and safe treatment for neuroleptic-induced sialorrhea. Psychopharmacology (Berl) 207:593–597
234. Storr M, Allescher HD, Rosch T, Born P, Weigert N, Classen M (2001) Treatment of symptomatic diffuse esophageal spasm by endoscopic injections of botulinum toxin: a prospective study with long-term follow-up. Gastrointest Endosc 54:754–759
235. Storr M, Born P, Frimberger E, Weigert N, Rösch T, Meining A, Classen M, Allescher HD (2002) Treatment of achalasia: the short-term response to botulinum toxin injection seems to be independent of any kind of pretreatment. BMC Gastroenterol 2:19
236. Strutton DR, Kowalski JW, Glaser DA, Stang PE (2004) US prevalence of hyperhidrosis and impact on individuals with axillary hyperhidrosis: results from a national survey. J Am Acad Dermatol 51:241–248
237. Suskind DL, Tilton A, Suskind DL, Tilton A (2002) Clinical study of botulinum-A toxin in the treatment of sialorrhea in children with cerebral palsy. Laryngoscope 112:73–81
238. Sycha T, Graninger M, Auff E, Schnider P (2004) Botulinum toxin in the treatment of Raynaud's phenomenon: a pilot study. Eur J Clin Invest 34:312–313
239. Tcherniak A, Kashtan DH, Melzer E (2006) Successful treatment of gastroparesis following total esophagectomy using botulinum toxin. Endoscopy 38:196
240. Thomson AJ, Jarvis SK, Lenart M, Abbott JA, Vancaillie TG (2005) The use of botulinum toxin type A (BOTOX) as treatment for intractable chronic pelvic pain associated with spasm of the levator ani muscles. BJOG 112:247–249

241. Tranqui P, Trottier DC, Victor C, Freeman JB (2006) Nonsurgical treatment of chronic anal fissure: nitroglycerin and dilatation versus nifedipine and botulinum toxin. Can J Surg 49: 41–55

242. Tsai SJ, Ying TH, Huang YH, Cheng JW, Bih LI, Lew HL (2009) Transperineal injection of botulinum toxin A for treatment of detrusor sphincter dyssynergia: localization with combined fluoroscopic and electromyographic guidance. Arch Phys Med Rehabil 90:832–836

243. Vadoud-Seyedi J, Heenen M, Simonart T (2001) Treatment of idiopathic palmar hyperhidrosis with botulinum toxin. Report of 23 cases and review of the literature. Dermatology 203: 318–321

244. Vadoud-Seyedi J (2004) Treatment of plantar hyperhidrosis with botulinum toxin type A. Int J Dermatol 43:969–971

245. Vaezi MF, Richter JE, Wilcox CM, Schroeder PL, Birgisson S, Slaughter RL, Koehler RE, Baker ME (1999). Botulinum toxin versus pneumatic dilatation in the treatment of achalasia: a randomized trial. Gut 44:231–239

246. Verma A, Steele J (2006) Botulinum toxin improves sialorrhea and quality of living in bulbar amyotrophic lateral sclerosis. Muscle Nerve 34:235–237

247. von Lindern JJ, Niederhagen B, Bergé S, Hägler G, Reich RH (2000) Frey syndrome: treatment with type A botulinum toxin. Cancer 89:1659–1663

248. Wabbels B, Förl M (2007) Botulinum toxin treatment for crocodile tears, spastic entropion and for dysthyroid upper eyelid retraction. Ophthalmologe 104:771–776

249. Wehrmann T, Seifert H, Seipp M, Lembcke B, Caspary WF (1998) Endoscopic injection of botulinum toxin for biliary sphincter of Oddi dysfunction. Endoscopy 30:702–707

250. Wehrmann T, Kokabpick H, Jacobi V, Seifert H, Lembcke B, Caspary WF (1999) Long-term results of endoscopic injections of botulinum toxin in elderly achalasic patients with tortuous megaesophagus or epiphrenic diverticulum. Endoscopy 31:352–358

251. Wehrmann T, Schmitt TH, Arndt A, Lembcke B, Caspary WF, Seifert H (2000) Endoscopic injection of botulinum toxin in patients with recurrent acute pancreatitis due to pancreatic sphincter of Oddi dysfunction. Aliment Pharmacol Ther 14:1469–1477

252. Wilken B, Aslami B, Backes H (2008) Successful treatment of drooling in children with neurological disorders with botulinum toxin A or B. Neuropediatrics 39:200–204

253. Wollina U (2008) Pharmacological sphincterotomy for chronic anal fissures by botulinum toxin A. J Cutan Aesthet Surg 1:58–63

254. Woodward MN, Spicer RD (2003) Intrapyloric botulinum toxin injection improves gastric emptying. J Pediatr Gastroenterol Nutr 37:201–202

255. Wu KP, Ke JY, Chen CY, Chen CL, Chou MY, Pei YC (2011) Botulinum toxin type A on oral health in treating sialorrhea in children with cerebral palsy: a randomized, double-blind, placebo-controlled study. J Child Neurol 26:838–843

256. Yamashita N, Shimizu H, Kawada M, Yanagishita T, Watanabe D, Tamada Y, Matsumoto Y (2008) Local injection of botulinum toxin A for palmar hyperhidrosis: usefulness and efficacy in relation to severity. J Dermatol 35:325–329

257. Zaninotto G, Annese V, Costantini M, Del Genio A, Costantino M, Epifani M, Gatto G, D'onofrio V, Benini L, Contini S, Molena D, Battaglia G, Tardio B, Andriulli A, Ancona E (2004) Randomized controlled trial of botulinum toxin versus laparoscopic heller myotomy for esophageal achalasia. Ann Surg 239:364–370

258. Zárate N, Mearin F, Wang XY, Hewlett B, Huizinga JD, Malagelada JR (2003) Severe idiopathic gastroparesis due to neuronal and interstitial cells of Cajal degeneration: pathological findings and management. Gut 52:966–970

259. Zhu Q, Liu J, Yang C (2009) Clinical study on combined therapy of botulinum toxin injection and small balloon dilation in patients with esophageal achalasia. Dig Surg 26:493–498

Chapter 5
Clinical Use of Botulinum Neurotoxin: Urogenital Disorders Including Overactive Bladder

Alex Gomelsky and Roger R. Dmochowski

Abstract Botulinum neurotoxin A (BoNT-A) has emerged as an alternative treatment for many disorders of the lower urologic tract. This toxin inhibits acetylcholine (ACh) release at the presynaptic cholinergic neuromuscular junction and induces a flaccid paralysis. This impact on the bladder has made BoNT-A an attractive therapeutic option for detrusor overactivity (DO) due to neurogenic and idiopathic reasons that are refractory to traditional treatment with muscarinic receptor antagonists. Likewise, the injection of BoNT-A into the prostate and the external urethral sphincter has been used for the treatment of benign prostatic hypertrophy (BPH) and detrusor external sphincter dyssynergia, respectively. The injection of BoNT-A may also have an antinociceptive effect as therapeutic benefit has also been shown in painful bladder conditions such as interstitial cystitis (IC). Outcomes in the pediatric population are encouraging, and serious adverse events after BoNT-A injection are infrequent.

Keywords Botulinum neurotoxin · Detrusor · Prostate · Bladder · Urothelium · Urologic · Cystitis

5.1 Introduction

Botulinum neurotoxin (BoNT) is produced by the gram-positive, rod-shaped anaerobic bacterium *Clostridium botulinum*. This bacterium was initially isolated by van Ermengem in 1897, and BoNT is considered by most researchers to be the most potent biological toxin known to man [1], [2]. The toxin acts by inhibiting acetylcholine (ACh) release at the presynaptic cholinergic neuromuscular junction [3],

R. R. Dmochowski (✉)
Department of Urologic Surgery, Vanderbilt University Medical Center,
A-1302, Medical Center North, Nashville TN 37232, USA
e-mail: roger.dmochowski@Vanderbilt.edu

A. Gomelsky
Department of Urology, Louisiana State University Health—Shreveport,
1501 Kings Highway, Shreveport LA 71130, USA
e-mail: agomel@lsuhsc.edu

K. A. Foster (ed.), *Clinical Applications of Botulinum Neurotoxin,*
Current Topics in Neurotoxicity 5, DOI 10.1007/978-1-4939-0261-3_5,
© Springer Science+Business Media New York 2014

and, in sufficient quantities, ingestion of BoNT may cause a variety of symptoms, from food poisoning (botulism) to paralysis and ultimately death [4]. BoNT has traditionally been used for the treatment of neural and muscle-related conditions, such as strabismus, focal dystonias, limb spasticity, and dysphagia ([5], Chapter 3 of this volume). In the past two decades, this therapeutic modality has also garnered significant attention in the urologic community owing to its effects on lower urinary tract end organs such as the bladder and prostate. The objectives of this chapter are to elaborate on the pathophysiology of BoNT in the urinary tract and to evaluate the impact and complications of this therapy on various urologic conditions.

5.2   Structure and Mechanism of Action

Seven immunologically distinct BoNT serotypes are produced by *C. botulinum*: A, B, C1, D, E, F, and G; however, only types A and B are used clinically. Type A, the more potent and the longer lasting, is the most commonly used serotype in urologic applications. BoNTs are commonly used clinically as a purified protein complex (see Chap. 4 of the companion volume to this book, KA Foster (ed) *Molecular Aspects of Botulinum Neurotoxin*, Springer, New York) that range in size from 300 to 900 kDa and include a commonly shared 150-kDa exotoxin that is responsible for the pharmacological activity and protective non-toxin non-hemagglutinin and hemagglutinin accessory proteins [6]. In the lower urinary tract, the end-organ effects of BoNTs are mediated at the parasympathetic presynaptic nerve terminal [3]. The toxin is produced within the bacterial cytosol, and proteolytic cleavage of the 150-kDa polypeptide results in a 100-kDa heavy chain and a 50-kDa light chain that remain linked by a non-covalent protein interaction and a disulfide bond [7]–[9]. The heavy chain is considered to be responsible for the toxin binding to its serotype-specific acceptors on the target cell, while the light chain actively cleaves a specific site on a protein complex responsible for docking and release of vesicles containing neurotransmitters from the neuron [10], [11].

To understand the mechanism of BoNT action, it is first important to review the normal release of ACh from the nerve terminal [3]. While this process is complex, there are several key steps. In the unactivated state, the presynaptic terminals of the parasympathetic nerves display syntaxin and synaptosomal-associated membrane protein (SNAP; 25 kDa) on the inner surface of their plasma membrane, while synaptic vesicles (SVs) containing ACh are located in the cytosol. Neurotransmitter release involves the adenosine-5′-triphosphate (ATP)-dependent transport of neurotransmitter-containing vesicles to the plasma membrane and release of the neurotransmitter. Following nerve activation, membrane depolarization, and intracellular calcium influx, synaptobrevin that is located on the SVs forms a soluble N-ethylmaleimide-sensitive factor attachment protein receptor (SNARE) complex of proteins with the SNAP-25 and syntaxin on the presynaptic membrane. Fusion of the vesicular membrane with the presynaptic cell membrane results in release of

ACh into the synaptic cleft, a process called exocytosis. Subsequently, post-junctional muscarinic receptors are activated and detrusor contraction occurs.

Four steps are required for BoNT-induced paralysis [3]. First, the neurotoxin diffuses to the cholinergic nerve terminals, where the heavy chain binds to a SV nerve terminal receptor (SV2 or synaptotagmin). Second, the toxin is internalized within the nerve terminal via receptor-mediated endocytosis. Next, the light chain moiety translocates from the endocytotic vesicle through the vesicle membrane via a protein channel formed by the amino-terminal half of the heavy chain, and the disulfide bond is reduced allowing the light chain to enter into the cytosol. Finally, inhibition of ACh release is achieved when the light chain moiety cleaves specific SNARE proteins, thus exerting its clinical effects. Particular SNARE proteins are targeted by specific clostridial neurotoxins. For example, BoNTs A and E specifically target SNAP-25 [12]. BoNT-C1 cleaves both SNAP-25 and syntaxin, while BoNTs B, D, F, and G cleave vesicle-associated membrane protein (VAMP)/synaptobrevin [13], [14]. Whichever SNARE protein is cleaved, the result is the same: Vesicle fusion with the presynaptic plasma membrane is blocked.

5.3   Pathophysiology of BoNT

BoNT exerts its effects by inhibiting ACh release at the neuromuscular junction and direct inhibition of ATP release from presynaptic nerve terminals. The end result is induction of a flaccid paralysis from reversible chemodenervation [11], [15]. Since nerve terminals affected by BoNT do not regenerate, their function typically recovers due to axonal sprouting and formation of novel synaptic connections [16]. Subsequently, clinical response wanes with time and there may be a need for repeated injections of BoNT. As new axons contain all of the components necessary for exocytosis, these axons have the capability to form functional sprouts [17], [18]. However, there is a well-recognized second phase where the sprouts eventually degenerate and synaptic activity resumes at the original nerve terminals [15]. Degeneration in bladder cholinergic nerves remains incompletely understood, as an ultrastructural study of human bladder specimens before and after BoNT-A injection was unable to detect either signs of nerve fiber degeneration or signs of nerve fiber sprouting [19]. The time to recover neuromuscular function after induction of paralysis varies significantly from 2–4 months to over a year [17], [20]. There is also emerging evidence that BoNT-A injection may affect sensory transmission and modulate the expression of several urothelial growth factors and peptides. BoNT-A reduces the release of glutamate and substance P from sensory neurons and also appears to reduce the release of neuropeptides such as calcitonin gene-related peptide (CGRP) from the peripheral terminals of afferent bladder neurons [21], [22]. BoNT-A injection is also associated with a decrease of sensory fibers immunoreactive to P2X3 and TRPV1 [23]. These findings suggest that BoNT-A may address bladder sensory disorders along several locations in the cascade [24].

5.4 Clinical Applications

Commercial preparations of BoNT have different doses, efficacy, and safety profiles and should not be considered interchangeable. In the USA, commercially available botulinum toxins include Botox® (type A; Allergan, Inc., Irvine, CA, USA; Food and Drug Administration (FDA)-recommended name: OnabotulinumtoxinA), Dysport® (type A; Ipsen Biopharm Ltd., Wrexham, UK; AbobotulinumtoxinA), and Myobloc® (type B; Elan Pharmaceuticals, Inc., San Francisco, CA, USA; RimabotulinumtoxinB). An additional type A formulation is available in Europe (Xeomin®; Merz Pharma GmbH, Frankfurt am Main, Germany). Doses of BoNT are represented by mouse units (U), with one unit of toxin representing the amount necessary to kill 50 % of a group of female mice. The extrapolated lethal dose of BoNT in humans would range from 2000 to 3000 units, and since the typical injection doses range from 200 to 300 units, systemic weakness or paralysis is unlikely even if BoNT is injected directly into the bloodstream [3]. For the remainder of the chapter, further discussion of BoNT-A will refer to the Botox®/OnabotulinumtoxinA formulation.

5.5 Injection Technique

In the urologic tract, BoNT-A is most commonly administered via injection into the detrusor, external sphincter, or prostate, and, more recently, via instillation into the bladder. The amount of toxin, dilution volume, location and number of injections are only some of the factors to be taken into consideration when planning intradetrusor injection. Additionally, these variables may change based on the indication for injection.

The injection procedure has recently evolved into a minimally invasive option with minimal anesthetic requirements and the ability to perform the procedure effectively in the office setting. Antibiotic prophylaxis should be used with intradetrusor BoNT-A injections, as the rate of symptomatic urinary tract infection (UTI) after BoNT-A injection may exceed 7 % [25]. The urinary colonization rate is 31 and 26 % at 6 days and 6 weeks after injection, respectively, with *Escherichia coli* being the most common bacterial pathogen isolated (62.5 % of the time). Prior to injection, anesthesia is delivered via instillation of 50 mL of 1 % lidocaine solution through a urethral catheter and left to dwell in the bladder for 15–30 min. Schurch et al. have also suggested that electromotive drug administration (EMDA) may enhance pain control as compared to intravesical lidocaine alone [26].

Delivery of BoNT-A may be achieved with a rigid or flexible cystoscope and a 25-gauge flexible Williams needle. Others have employed a collagen injection needle (Bard Medical, Covington, GA, USA). Most studies have reported injection doses of 100–300 U of BoNT-A, with 100 U being a typical starting dose for idiopathic detrusor overactivity (IDO) and higher doses for those patients with insufficient response to an initial injection or those with neurogenic detrusor overactivity (NDO). The 100–300 U of BoNT-A are diluted in normal saline at a concentration of 10 U per mL. The solution is injected into the detrusor in 0.5–1.0 mL doses at 10–30 sites in a grid-like pattern or in thirds, on the right, middle, and left portions of the posterior and lateral walls of the bladder. The injection depth is typically submucosal or superficially intramuscular, and the objective is to raise a submucosal bleb without blanching the urothelium. The solution can be combined with 0.1 mL of indigo carmine to allow a more precise mapping of the injection pattern. The decision on whether to inject or spare the trigone has been widely debated, owing to the concern of acute urinary retention (AUR) and vesicoureteral reflux (VUR) after trigonal injection. Certainly, for patients with NDO already performing clean intermittent catheterization (CIC), this has not been of significant concern. However, there remains a risk of AUR after trigonal injection in patients with IDO, and most studies have reported outcomes in this patient population after trigone-sparing BoNT injections.

The technique varies slightly for injection of the external urethral sphincter, with 200 U of BoNT-A diluted in 4 mL of normal saline. In men, 1 mL injections are performed in the external sphincter with a Williams needle at the 3, 6, 9, and 12 o'clock positions, while in women, periurethral injection is performed using a 22-gauge spinal needle at the 3, 6, 9, and 12 o'clock positions. The suggested depth of injection is 2 cm and approximately 10 mm parallel to the urethra.

Prostate injections may be performed via the transperineal, transurethral, and transrectal approaches, with 100–300 U of BoNT-A delivered in volumes between 4 and 20 mL [27]. The amount of BoNT-A delivered varies with prostate size. Patients with a prostate size of less than 30 mL typically receive 100 U and men with larger prostates receive 200 U [27]. Men with prostate volumes exceeding 60 mL may require more than 200 U. Transrectal ultrasound (TRUS) guidance has been used to direct a 21-gauge, 20-cm Chiba needle through the perineum while the man is in the lithotomy position. The needle is inserted 1 cm to the left and 1 cm to the right of the median raphe, and 1–3 cm above the anal sphincter [27]–[29]. The transverse view has been used to ensure proper needle placement as a bright spot in the center of the transition zone. The scanning plane is changed to longitudinal and the needle is further advanced until it is 0.5–1.0 cm from the bladder neck. BoNT-A is then injected at the cranial, middle, and caudal aspects of the lateral lobe with diffusion over the lateral prostate lobe confirmed by TRUS monitoring [27].

5.6 Clinical Outcomes of BoNT-A Injection for Urologic Conditions

As with many other disease processes, studies reporting outcomes are often difficult to compare. Outcomes after BoNT-A administration may be reported as objective measures culled from a voiding diary or urodynamic assessment or subjective measures, obtained from responses to validated quality of life (QOL) questionnaires and indices. The most common measures are summarized in Table 5.1.

Table 5.1 Outcome measures commonly reported in BoNT-A studies

Measure	Abbreviation	Units
Voiding diary characteristics		
Incontinence episodes	IE	#/day
Urinary frequency	UF	#/day
Total voids	TV	#/day
Nocturnal voids	NV	#/day
Urinary urgency	UU	#/day
Urgency urinary incontinence	UUI	#/day
Bladder capacity	BC	mL
Quality of life (QOL) indices		
Quality of life score	QOL	
Urogenital distress inventory	UDI-6	
Incontinence impact questionnaire	IIQ-7	
Patient global impression of improvement	PGI-I	
King's Health questionnaire	KHQ	
International prostate symptom score	IPSS	
International quality of life questionnaire	I-QOL	
Urodynamic indices		
Maximum cystometric capacity	MCC	mL
Maximal urinary flow rate	Q_{max}	mL/sec
Volume at first void	V_{1V}	mL
Volume at urgency	V_U	mL
Volume at leakage	V_L	mL
Maximal detrusor pressure	P_{detmax}	cm H_2O
Detrusor pressure at maximal flow	P_{det} @ Q_{max}	cm H_2O
Bladder reflex volume	BRV	mL
Mean voiding pressure	MVP	cm H_2O
Post void residual	PVR	mL
Others		
Pad use	Pads	#/day
Pad weight	PW	g/day
Duration of response	DR	months

5.6.1 Idiopathic Detrusor Overactivity (IDO)

The first mention of BoNT-A injection for non-neurogenic urinary storage disorders was by Radziszewski et al. in 2001, with the authors reporting resolution of urinary incontinence after detrusor injections in seven patients participating in a pilot study [30]. The data regarding IDO and overactive bladder (OAB) symptoms refractory to medical therapy with muscarinic receptor antagonists and behavioral therapy continue to emerge, and, in January of 2013, the US FDA approved the use of onabotulinumtoxinA for IDO and refractory OAB.

A large, phase II, dose–response study evaluating BoNT-A for IDO investigated the optimal injection dose in this patient population [31]. This was a multinational, randomized, placebo-controlled, and parallel group study that enrolled 313 patients with idiopathic OAB that was not adequately managed with muscarinic receptor antagonists (Table 5.2). Patients reported more than eight weekly urgency urinary incontinence (UUI) episodes and had an average of eight or more micturitions per day. The presence of urodynamic DO was not required. Patients received a single treatment of 50, 100, 150, 200, or 300 U BoNT-A or placebo. The authors found that BoNT-A demonstrated significant dose- dependent improvements in urinary symptoms when compared with placebo. Furthermore, urodynamic parameters in patients with idiopathic OAB and improvement in symptoms were reflected in the patient's perception of treatment benefit [32]. Doses above 150 U did not appear to add much incremental benefit, particularly when balanced with post void residual (PVR)-related safety parameters.

Tincello et al. performed a double blind, placebo-controlled, randomized trial in eight UK urogynecology centers between 2006 and 2009 [33]. A total of 240 women with refractory DO were randomized to active or placebo treatment and followed up for 6 months. The treatment consisted of 200 U BoNT-A or placebo injected into the detrusor in 20 sites with the primary outcome being voiding frequency per 24 h. Median voiding frequency was significantly lower after BoNT-A compared with placebo (8.3 vs. 9.67), and similar differences were seen in urgency episodes (3.83 vs. 6.33) and leakage episodes (1.67 vs. 6.0). Continence was more common after BoNT-A (31 vs. 12 %). UTI (31 vs. 11 %) and voiding difficulty requiring CIC (16 vs. 4 %) were significantly more common after BoNT-A.

Recently, the results of two double blind, placebo-controlled, randomized, multicenter, phase III, 24-week clinical studies were reported [34], [35]. A total of 1,105 patients inadequately managed with antimuscarinic therapy were randomized to receive 100 U of BoNT-A ($n = 557$) or placebo ($n = 548$). At week 12, patients receiving BoNT-A in both studies demonstrated significant improvements versus placebo in all primary outcome variables (daily frequency of urinary incontinence episodes (IEs), micturition episodes, and voided volume (VV)). The adverse reactions seen in ≥ 2 % of patients in the BoNT-A group versus placebo were as follows: UTI (18 vs. 6 %), dysuria (9 vs. 7 %), urinary retention (6 vs. 0 %), bacteriuria (4 vs. 2 %), and elevated PVR not requiring CIC (3 vs. 0 %). The proportion of patients undergoing BoNT-A requiring CIC at any time during the complete treatment cycle was higher versus those undergoing placebo (6.5 vs. 0.4 %). The duration of CIC was for a median of 63 days versus 11 days for patients after placebo injection.

A subanalysis of the trial by Dmochowski et al. [31] revealed that a single BoNT-A treatment (≥ 100 U) resulted in statistically significant and clinically meaningful improvement in health-related QOL (HRQOL) as measured by the international quality of life questionnaire (I-QOL), the King's Health questionnaire (KHQ) symptom component and the Medical Outcomes Study 36-Item Short-Form Health Survey by week 2 compared with placebo and sustained for up to 36 weeks [36]. Assessment with a modified OAB-Patient Satisfaction with Treatment Questionnaire (PSTQ) and four Patient Global Assessment questions that assessed changes in symptoms, QOL, activity limitations and emotions also showed that patients with OAB are more likely to be satisfied and/or achieve their primary treatment goal with BoNT-A treatment than with placebo [37].

Table 5.2 Outcomes of BoNT-A injection in IDO (Idiopathic Detrusor Overactivity)

Author	Year	Trial	N	Dose	Outcomes
Zermann [31]	2001	Single center observational	7	50–200 U	(Trigone included); 57.1 % ↓ UF, ↑ BC
Rapp [32]	2004	Single center observational	35	300 U	34 % symptom resolution, 26 % slight improvement on IIQ-7, UDI-6
Schulte-Baukloh [33][a]	2005	Single center observational	44	200–300 U	@ 4w, 3m, 6m, 9m: ↓ UF 12 %, 16 %, 13 %, 9 %; ↓ pads/d @ 6m; @ 4w: ↑ BRV, ↑ MCC; 86 % would choose procedure again
Kuo [34]	2005	Prospective	20	200 U	45 % continent, 40 % improved @ 3m; 35 % continent @ 6m
Kessler [35][c]	2005	Prospective	22	300 U	Significant ↓ TV, ↓ NV, ↓ pads/d, ↑ MCC, ↓ P_{det} @Q_{max}
Popat [36][c]	2005	Prospective open label	75	200 and 300 U	Significant ↑ MCC, ↑ P_{det} @Q_{max}, ↓ TV, ↓ IE, ↓ UU; outcomes similar in IDO & NDO groups
Werner [37]	2005	Prospective single center	26	100 U	60 % improved/success; 35 % ↓ UF, 58 % ↑ MCC, 60 % ↓ DO
Schmid [38]	2006	Multicenter prospective open-label	100	100 U	92 % responders; ↑ MCC by 56 %; DO resolved in 74 %; ↑ V1v, ↑ Vu; ↓ UU, ↓ UF, ↓ NV, ↓ UUI; KHQ domains significantly improved @ 4w, 12w
Sahai [39]	2007	Single center RCT	16	200 U	@ 4w, 12w: significant ↑ MCC, ↓ UF, ↓ IE, ↓ IIQ-7, ↓ UDI-6; 10m DR
Mohanty [40]	2008	Observational cohort	35	200 U	Within 1w: 85 % improvement in UF, UU, NV, UUI; ↑ MCC, ↓ P_{det} by 49 %, ↑ V_{1v}; 7m DR
Brubaker [41]	2008	Multicenter RCT	28	200 U	373d median DR; significant ↓ PGI-I @ 2m (60 %), ↓ IE, ↓ UDI-6 (urge)
White [42][c]	2008	Observational cohort	21	200U	76 % responders; @ 1m: @ 50 % ↓ TV, ↓ pads/d; 7.12m DR
Sahai [43][b]	2009	Single Center RCT	16	200U	KHQ subdomains significantly improved in treatment group @ 4w, 12w
Flynn [44]	2009	Single center RCT	15	200 and 300 U	@ 6w: significant ↓ IE, ↓ IIQ-7, ↓ UDI-6, ↓ PW, ↓ pads/d; No △ in TV, NV, MCC, Q_{max}, P_{det}
Khan [45]	2010	Prospective open label	74	200 U	@ 4w: 51 % complete continence (per UDI-6); in those not cured, significant improvement in UU, SUI, and UF subscores of UDI-6
Kuo [46][c]	2010	Observational cohort	217		66.3 % successful outcome
Lie [47]	2010	Prospective single center	19	200 U	@ 3m: significant improvement in QOL, IE, BC, void interval; significant ↓ BRV, ↓ P_{det} max, ↓ V_L

RCT randomized, placebo-controlled trial

[a]22 patients also received external sphincter injections

[b]Follow-up evaluation of Sahai et al. [39]

[c]Included patients with IDO and NDO

Several authors have suggested risk factors for treatment failure in this population. Schmid et al. identified low baseline bladder compliance and maximum cystometric capacity (MCC) < 100 mL as risk factors for treatment failure, while Sahai et al. suggested that maximal detrusor pressure (P_{detmax}) > 110 cm H_2O may predict a poor response to 200 U injection [38], [39]. Higher doses may be necessary in this population. Additionally, in a retrospective analysis of 85 patients who underwent BoNT-A injections for IDO, Makovey et al. determined that BoNT-A therapy may be more successful in patients with anticholinergic intolerability as compared to patients with poor medication efficacy (86 vs. 60 %, $P = 0.02$) [40].

Response after the first intradetrusor BoNT-A injection may be seen very quickly. Khan et al. reported that the patient-reported outcome of complete continence (defined as a score of 0 in both the urgency and stress urinary incontinence (SUI) subscales of the urogenital distress inventory, UDI-6) was 51 % 4 weeks after the first BoNT-A 200 U injection in 74 patients [41]. In those who did not achieve continence, the median urgency, frequency, and SUI subscores decreased significantly from pretreatment values.

Several groups have evaluated the safety and benefits of BoNT-A reinjection for the treatment of IDO. Sahai et al. reinjected 20 of their original 34 patients with 200 U, and nine subsequently underwent a third and fourth injection [42]. Significant improvement was observed in OAB symptoms and QOL measures after each injection. MCC and compliance increased with a decrease in P_{detmax} during filling cystometography (CMG). When comparing OAB symptoms, QOL, and urodynamic parameters 3 months after the first and last injections, no significant differences were found. Studies by Khan et al. and Game et al. noted that the mean UDI-6 and incontinence impact questionnaire (IIQ-7) scores decreased after each reinjection, with a median interval between reinjections typically being 6–12 months [43], [44]. Likewise, additional studies by Granese et al. and Dowson et al. confirmed that reinjections are safe and efficacious [45], [46]. Finally, a recent review concluded that reinjections were safe and that reproducible and sustained improvement in symptoms may be seen after each injection [47].

In 2012, the American Urological Association listed BoNT-A injection as a third-line treatment option for OAB for those patients refractory to first-line (behavioral therapy) and second-line (antimuscarinic medications) interventions (Evidence Strength Grade C) [48].

5.6.2 *Neurogenic Detrusor Overactivity (NDO)*

In 2000, Schurch et al. first reported on the outcomes of BoNT-A in the treatment of patients with refractory NDO [49], [50]. Nineteen of 21 patients enrolled in a prospective, nonrandomized study were evaluated at 6 weeks after BoNT-A injection and 11 were evaluated at 16 and 36 weeks. Complete continence was restored in 17 of 19 cases in which the dose of antimuscarinic medication was markedly decreased or eliminated altogether. Significant improvement was seen in bladder reflex volume (BRV), MCC and maximum detrusor voiding pressure. Mean PVR increased significantly after injection; however, autonomic dysreflexia manifesting as a hypertensive crisis during voiding disappeared in three patients with tetraplegia. Satisfaction was high in the entire group and the overall improvement in urodynamic parameters and incontinence was persistent in those patients evaluated at 16 and 36 weeks. An updated series by Schurch et al. described the outcomes of a multicenter evaluation of BoNT-A at eight European sites [51]. Fifty-nine patients with NDO mostly due to spinal cord injury (SCI) were enrolled and received a single injection of 200–300 U BoNT-A with subsequent assessments at up to 6 months. There were significant posttreatment decreases in IEs from baseline in the two BoNT-A groups but not in the placebo group. In addition, more patients who received BoNT-A reported no IEs during at least one posttreatment evaluation period. Positive treatment effects were also reflected by significant improvements in bladder function in the BoNT-A groups, as assessed by urodynamics and in patient QOL. Benefits were observed from the first evaluation at 2 weeks to the end of the 24-week study. No safety concerns were raised. Additional studies have likewise demonstrated the beneficial effects of BoNT-A in NDO (Table 5.3). The predominant indication for BoNT-A injection has been suboptimal results obtained with antimuscarinic medications. The US FDA approved the use of BoNT-A for NDO in 2011.

Several prospective, randomized trials evaluating BoNT-A for NDO have since been published. In a prospective, double blind, multicenter study, patients with NDO due to SCI and multiple sclerosis (MS) and ≥ 1 daily episodes of urinary incontinence despite current antimuscarinic treatment were randomized to BoNT-A 300 U ($n = 28$) or placebo ($n = 29$) [70]. Injection was performed at 30 intradetrusor sites and the trigone was spared. Patients were offered open-label BoNT-A 300 U at week 36 and followed up an additional 6 months while 24 patients, each in the treatment and placebo groups, received open-label therapy. The primary efficacy parameter was daily urinary incontinence frequency at week 6, while secondary parameters were changes in the International Consultation on Incontinence Questionnaire and the urinary incontinence QOL scale at week 6. The mean daily frequency of UI episodes was significantly lower for BoNT-A than for placebo at week 6 (1.31 vs. 4.76), and for weeks 24 and 36. Improved urodynamic and QOL indices for treatment versus placebo were evident at week 6 and persisted to weeks 24–36. The most common adverse event in each group was UTI.

Another multicenter, double blind, randomized, control trial (RCT) enrolled patients with MS or SCI who had ≥ 14 UI episodes per week pretreatment [71]. Patients received 30 trigone-sparing injections of BoNT-A: 200 U ($n = 92$), 300 U ($n = 91$) or placebo ($n = 92$). At week 6, BoNT-A of 200 and 300 U significantly reduced weekly UI episodes compared with placebo, and benefit with BoNT-A was observed by the first posttreatment visit at week 2. Improvements in MCC, P_{detmax} at first involuntary detrusor contraction and I-QOL at 6 weeks were significantly greater with both BoNT-A doses compared to placebo. Benefits were observed in both the MS and SCI populations. The median time to patient request for retreatment was the same for both BoNT-A doses (42.1 weeks) and significantly greater than placebo (13.1 weeks). Significant increases in PVR were observed in patients not using CIC

Table 5.3 Outcomes of BoNT-A injection in NDO (Neurogenic Detrusor Overactivity)

Author	Year	N	Cause	Dose	Outcomes
Reitz [52]	2004	231		300 U	@ 12w, 36w: significant ↑ MCC, ↑ BRV; ↓ MVP; @ 12w: significant ↑ compliance; most patients reduced or stopped antimuscarinics
Bagi [53]	2004	15	SCI	300 U	87 % fully continent; significant ↓ P_{det} max, ↑ V_{max} @ $P_{det} < 40$ cm H_2O; ↑ MCC; 4–12m DR
Kessler [54][e]	2005	22	[f]	300 U	Significant ↓ TV, ↓ NV, ↓ pads/d, ↑ MCC, ↓ P_{det} @Q_{max}
Popat [55][e]	2005	75		200 & 300 U	Significant ↑ MCC, ↑ P_{det} @Q_{max}, ↓ TV, ↓ IE, ↓ UU; outcomes similar in IDO & NDO groups
Hajebrahimi [56]	2005	10	SCI	300–400 U	@ 3m: significant ↑ BRV, ↑ MCC, ↓ P_{det} max
Klaphajone [57]	2005	10	SCI	300 U	(low bladder compliance population); @ 6w: significant ↑ compliance, ↑ BRV, ↓ P_{det} max ↑ BC; all urodynamic variables significantly improved @ 16w, but returned to baseline by 36w
Kuo [58]	2006	24	CVA, SCI	200 U	@ 1m: ↑ BC, ↑ BRV; P_{det} ↓ more for SCI pts; complete continence/improvement in 8.3/41.7 % of cerebrovascular accident (CVA) pts & 33.3/58.3 % SCI pts; all symptoms relapsed by 6m
Schulte-Baukloh [59]	2006	11	MS	300 U	@ 4w: ↑ PVR (CIC 1 pt): @ 4w, 3m, 6m: ↓ UF, ↓ NV, ↓ pads/d (4w, 3m); ↑ BRV, ↑ MCC, ↓ MDP; Subjective improvement @ 4w, 3m
Tow [60]	2007	15	SCI	300 U	@ 6w, 26w: significant ↓ IE; 75 % completely dry @ 6w, 50 % @ 39w; significant ↑ PVR, ↑ MCC, ↓ P_{det} max, ↑ BRV; satisfaction levels increased posttreatment
Stoehrer [61]	2009	216		300 U[c]	↑ MCC, ↑ BRV, ↑ compliance, ↓ P_{det} max; 9m DR; incontinence significantly improved
Giannantoni [62]	2009	6	PD, MSA	200 U	@ 1m, 3m: all ↓ UF, ↓ NV, improved QOL scores; by 5m all UU, UUI resolved; all ↑ PVR; all urodynamic parameters improved; CIC only in MSA pts
Alvares [63]	2010	22	SCI	300 U	↑ MCC, ↑ BRV, ↑ compliance, ↓ P_{det} max; 7m DR; 82 % required anticholinergics to achieve continence; 23 % had repeat injection; 41 % proceeded to augmentation
Giannantoni [64]	2010	17	SCI	300 U	@ 6y: significant ↓ IE, ↑ BRV, ↑ MCC, ↓ P_{det} of UDC; 88 % completely continent; QOL index significantly improved
Wefer [65]	2010	214	SCI[d]		Significant ↑ MCC, ↑ compliance, ↓ P_{det} max; significant ↓ UTIs, incontinence, incontinence aids

Table 5.3 (contined)

Author	Year	N	Cause	Dose	Outcomes
Mehnert [66]	2010	12	MS	100 U	Significant ↑ MCC, ↓ P_{det} max; significant ↓ UF, ↓ UU, ↓ pads/d; ↑ PVR initially; 8m DR
Gomes [67]	2010	42[a]	SCI[b]	200–300 U	↑ MCC, ↑ compliance, ↓ P_{det} max; continence achieved in 62 % by 12w
Chen [68]	2011	108	SCI	300 U	Significant ↑ MCC, ↑ BRV; @ 6w, 12w, 36w: significant ↓ MVP; I-QOL, satisfaction significantly improved
Chen [69]	2011	38	SCI	200 U	60 % satisfactory response; UDI-6, QOL significantly improved; significant ↑ MCC, ↑ PVR

BC bladder capacity, *BRV* bladder reflex volume, *CIC* clean intermittent catheterization, *DR* duration of response, *IDO* idiopathic detrusor overactivity, *IE* incontinence episodes, *I-QOL* international quality of life questionnaire, *MCC* maximum cystometric capacity, *MDP* maximum detrusor pressure, *MS* multiple sclerosis, *MSA* multiple system atrophy, *NDO* neurogenic detrusor overactivity, *NV* nocturnal voids, *PD* Parkinson's disease, *pts* patients, *PVR* post void residual, P_{detmax} maximal detrusor pressure, *SCI* spinal cord injury; *UDC* uninhibited detrusor contraction, *UDI-6* urogenital distress inventory, *UF* urinary frequency, *UTI* urinary tract infection, *UU* urinary urgency, *UUI* urgency urinary incontinence, *QOL* quality of life

[a] Some patients underwent Chinese BoNT-A (Prosigne®, Lanzhou Biological Products Institute, Lanzhou, China)

[b] 80 % SCI; remainder from viral myelitis, MS, and schistosomal myeloradiculopathy

[c] Some patients also received Dysport 750U

[d] 81 % SCI, 14 % myelomeningocele, 5 % MS

[e] Included patients with IDO and NDO

[f] SCI (2), traumatic brain injury (1), myotonic dystrophy (1), Friedrich's ataxia (1), PD (1), tethered cord (2), MS (3)

prior to treatment, and 12, 30, and 42 % of patients in the placebo, 200 U, and 300 U groups, respectively, initiated CIC after treatment. A subanalysis of this study revealed that NDO patients treated with BoNT-A 200 U or 300 U had significantly greater improvement in HRQOL and greater treatment satisfaction compared with placebo-treated patients, with no clinically relevant differences between BoNT-A doses [72].

Kennelly et al. evaluated the long-term efficacy and safety of repeat BoNT-A injections in patients with NDO inadequately managed by antimuscarinics [73]. Patients who completed either of the two preceding phase III studies were offered entry into an extension study and received repeat BoNT-A 200 U or 300 U. The primary assessment was the change from baseline in weekly UI episodes at 6 weeks after each treatment. Additional assessments included ≥ 50 and 100 % reductions in UI episodes, voided volume (VV), and I-QOL responses. A total of 387, 336, 241, 113, and 46 patients received 1, 2, 3, 4 and 5 BoNT-A treatments, respectively. Weekly UI episodes were consistently reduced compared with baseline after repeated BoNT-A treatment ($- 22.7, - 23.3, - 23.1, - 25.3$ and $- 31.9$ for the 200 U group in cycles 1–5). The proportion of patients reporting ≥ 50 and 100 % ("dry") reductions from baseline in UI episodes at week 6 ranged from 73 to 94 % and 36 to 55 %, respectively. Increases in the voided volume (VV) (mean increase $> 130\,\text{mL}$) and improvements in QOL were also observed after each repeat treatment. The most common adverse events were UTIs and urinary retention, with no change in the adverse event profile over time.

Overall satisfaction with BoNT-A injections in the NDO population appears to be high. In a 5-minute telephone interview of 72 patients with NDO, 66.7 % were still actively undergoing repeat BoNT-A injections [74]. Of these patients, 90 % replied that they would consider continuing with BoNT-A injections as a long-term treatment option, while only 15 % of those still undergoing injections thought that they would consider an alternative permanent surgical option in the next 5 years. Furthermore, intradetrusor injections of BoNT-A produce comparable and significant improvements in QOL at least 16 weeks (or more) after treatment [75]. This appears to hold true for patients with either NDO or IDO. Changes in lower urinary tract symptoms (LUTS), as opposed to improvements in urodynamic parameters, appear to be the major determinants of improvement in the QOL. Similarly, Schurch et al. found that the median total subscale I-QOL scores increased significantly from screening with BoNT-A 300 U compared with placebo at all time points (2, 6, 12, 18 and 24 weeks) and with BoNT-A 200 U compared with placebo at all time points for total score and the Avoidance Limiting Behavior subscale, and at weeks 2, 6, 12 and 18 for the Psychosocial Impact and Social Embarrassment subscales [76].

Karsenty et al. performed a systematic literature review encompassing 18 articles that evaluated the efficacy and safety of BoNT-A in patients with NDO [77]. The typical injected dose was 300 U and most studies reported a significant improvement in clinical (~ 40–80 % of patients became completely dry in between performing CIC) as well as urodynamic (in most studies mean P_{detmax} was reduced to $\leq 40\,\text{cm}\ H_2O$) variables and patients' QOL, without major adverse events. An additional evidence-based review conducted by the American Academy of Neurology concluded on the basis of two class I studies that BoNT-A should be offered as a treatment option for NDO (Level A evidence) [5].

The long-term effects of repeated BoNT-A injections have also been recently evaluated. Several groups concluded that reinjections of BoNT-A were effective, with improvement in urodynamic indices and continence status seen after each injection [78]–[80]. Most patients were able to eliminate or decrease their use of antimuscarinic medications. However, Pannek et al. reported some different outcomes from 27 consecutive patients with NDO who received at least five injections (mean 7.1 injections) [81]. After the first injection, MCC, BRV, continence status, and compliance were all significantly improved, and P_{detmax} was significantly reduced. Compared with the results after the first treatment, the incontinence rate and the number of patients with an elevated P_{detmax} were slightly increased after the final BoNT-A treatment. The long-term success rate was 74 %; however, every fourth patient required major surgical intervention. Additionally, there was a significant decrease in P_{detmax} before BoNT-A treatments, suggesting that BoNT-A may lead to impaired detrusor contraction strength that does not completely recover after treatment.

There has been a lot of debate as to whether to exclude the trigone in the injection of BoNT-A. While VUR has been traditionally viewed as a possible sequela of trigonal injection, studies by Karsenty et al. and Mascarenhas et al. have concluded that trigonal injection does not lead to VUR [82], [83]. Recent evidence has also emerged to suggest that inclusion of the trigone may be associated with better outcomes. Abdel-Meguid randomized 36 patients with SCI who discontinued anticholinergics to receive 300 U BoNT-A into the detrusor only or 200 U into the detrusor and 100 U into the trigone (combined arm) in a 1:1 ratio [84]. On within-group analysis, all parameters improved significantly compared to baseline. On between-group analysis at week 8, patients in the combined arm had a significantly greater decrease in incontinence (80.9 vs. 52.4 %) and complete dryness (66. 7 vs. 33.3 %). There was also a significant improvement in absolute difference for maximum bladder reflex volume (MBRV) in favor of the combined arm. Additionally, at week 18, anticholinergics were needed again in fewer patients in the combined arm.

There is also evidence that BoNT-A injections may lead to a decrease in symptomatic UTIs in patients with NDO. After injection in 30 patients, the mean number of symptomatic UTIs over 6 months decreased significantly from 1.75 ± 1.87 to 0.2 ± 0.41, and only three patients presented with symptomatic UTIs [85]. These patients were the ones who showed less improvement in their urodynamic parameters (e.g., BRV, MCC, P_{detmax}) after injection.

In conclusion, BoNT-A may be a logical choice for NDO refractory to therapy with muscarinic receptor antagonists. NDO stemming from various etiologies, including SCI, MS, Parkinson's disease (PD), and multiple system atrophy (MSA) appears to respond to BoNT-A injection. The optimal dose for patients with NDO appears to be 300 U. Therapeutic effects are often seen as early as 2 weeks after injection and the duration of effect is typically 6–12 months. There is a significant improvement in continence status and many patients are able to decrease or eliminate the use of antimuscarinics. Urodynamic parameters, such as MCC and P_{detmax}, are significantly improved and QOL scores are markedly improved. Repeat injections are typically associated with similar outcomes as the first injection. Trigonal inclusion is not associated with VUR and may actually improve outcomes.

5.6.3   Safety and Adverse Events Associated with BoNT-A

Adverse events associated with the act of injection, rather than BoNT-A itself, include bladder pain, hematuria and UTIs. As the mechanism of BoNT-A is local, these adverse events are typically mild and transient. Additionally, local effects from detrusor relaxation may include transient elevation in PVR and possibly AUR. A review of BoNT-A trials for IDO has revealed that urinary retention occurs in approximately 5 % of patients, while the need for CIC may approach 16 % [86]. However, as Nitti pointed out, the definitions of both "urinary retention" and the threshold PVR volume for beginning CIC are inconsistently defined in the literature. For example, Kuo et al. performed a multivariate analysis in 217 patients with IDO receiving their first injection [87]. Male gender and baseline PVR $\geq$ 100 mL were predictors of AUR, while baseline PVR $>$ 100 mL and injections of $>$ 100 U of BoNT-A were predictors of straining to void. The incidence of elevated PVR after the procedure was also associated with comorbidity and UTIs occurred more frequently in women and in men with a retained prostate. On the other hand, when AUR was defined as a PVR $>$ 200 mL irrespective of symptoms, 43 % of the patients in the study by Brubaker et al. satisfied the criteria [88]. The authors stopped their well-designed RCT after 75 % of their patients required antibiotic therapy for a UTI and all patients were started on CIC whether they were symptomatic or not. The median time to initiation of CIC was 30 days and the period of CIC lasted a median of 62 days. Patients requiring CIC after the first BoNT-A injection may also be more likely to require CIC with reinjections. Khan et al. defined a PVR $>$ 100 mL in association with LUTS as the indication to initiate CIC in patients treated with 200 U for IDO [43]. Of their 81 consecutive patients, 43 % required CIC after injection and there was no significant difference in the rate of CIC in patients who had repeated injections. The authors concluded that the possibility of AUR and a willingness to perform CIC should be mandatory in the informed consent process for patients considering this therapy.

Of interest, performing CIC does not appear to significantly impact a patient's QOL following BoNT-A injection for refractory IDO [89]. In the 43 % of 65 women requiring post-procedure CIC, the mean improvement in UDI-6 and IIQ-7 scores was not statistically different from those women not requiring CIC. QOL comparisons, likewise, did not reveal any significant differences between women performing CIC and those who did not. Thus, while all patients should be warned about the potential for CIC, the improvement in QOL appears to be similar to those women not requiring CIC. At least in this short-term study, the trouble of performing CIC did not outweigh the benefits of BoNT-A therapy.

The only absolute contraindications to BoNT-A injection are the presence of active infection and a known hypersensitivity to the agent. Relative contraindications include preexisting neuromuscular disorders (e.g., myasthenia gravis and Eaton–Lambert syndrome) and concomitant use of agents that may interfere with neuromuscular transmission (e.g., aminoglycosides and curare- like compounds), pregnancy (class C) and nursing mothers, and presence of bladder outlet obstruction. Since the mechanism of action of BoNT-A is local, adverse events are mild and transient and typically occur within the first week after injection. Systemic adverse events may occur if BoNT-A migrates away from the detrusor. The most common general adverse events are localized pain, tenderness and bruising, flu-like syndrome and muscle weakness. Rare adverse events include skin rash, pruritus, allergic reaction, dry mouth and dysphagia.

A boxed warning has been added to the prescribing information of BoNT-A in the USA to highlight the fact that BoNT may spread from the area of injection to produce symptoms consistent with botulism. Although rare, symptoms such as unexpected loss of strength or muscle weakness, hoarseness or trouble talking (dysphonia), trouble saying words clearly (dysarthria), loss of bladder control, trouble breathing, trouble swallowing, double vision, blurred vision and drooping eyelids may occur [86], [90], [91]. It is important to understand that swallowing and breathing difficulties can be life- threatening and there have been reports of deaths. It is also important to be aware that children being treated for spasticity are at greatest risk for severe adverse events. Finally, as previously stated, it is important to understand that BoNT products are not interchangeable and that the established drug names of the BoNT products have been changed. Naumann and Jankovic assessed the long- term safety of BoNT-A in a meta- analysis of 36 long- term studies encompassing 2309 patients [92]. Mild or moderate adverse events were reported in 25 % of patients receiving BoNT-A compared to 15 % of control patients. Focal weakness was the only adverse event that occurred significantly more often after BoNT-A treatment than control.

5.7   Other Urological Clinical Applications

5.7.1   Detrusor Sphincter Dyssynergia

The use of BoNT for detrusor sphincter dyssynergia (DSD) was actually the initial application in the urological field [93]. Ten of 11 men with SCI and DSD showed signs of sphincter denervation on electromyography and bulbosphincteric reflexes showed a decreased amplitude and normal latency. PVR decreased by an average of 146 mL after injection in eight patients and autonomic dysreflexia decreased in five patients. In the eight patients for whom it could be determined, toxin effects lasted for an average of 50 days. In a prospective study, Schurch et al. significantly improved DSD in 21 of 24 patients, with a concomitant decrease in PVR [94]. The effects of BoNT-A injection in this study lasted 3–9 months, necessitating reinjection. Additional studies have used varying injection techniques (transurethral vs. transperineal) and dosages of BoNT-A (50–240 U), with maximum duration of response equaling 13 months [3], [17]. Furthermore, there is evidence that sphincteric injection may improve dysfunctional voiding [3], [17].

An evidence-based review conducted by the American Academy of Neurology concluded on the basis of one class I study and two class II studies that BoNT-A should be considered for DSD after SCI (Level B) [5]. In the class I study, BoNT-A was compared to placebo as a single transperineal injection of 100 U in 86 patients with MS [95]. PVR was not decreased in the MS patients, which was different from typical results in patients with SCI. Two small class II studies found that BoNT-A injections for DSD associated with SCI were superior to placebo in terms of PVR, urethral pressure measurements and detrusor pressure variables (e.g., P_{detmax}) [96], [97]. Mild generalized weakness lasting 2–3 weeks was reported in three patients in the first study, while no significant adverse events were reported in the second study.

While the overwhelming majority of patients undergoing BoNT-A injection for DSD experience a significant decrease in PVR, treatment failure is often attributed to detrusor underactivity and bladder neck obstruction/dyssynergia [94], [98]. De novo SUI has also been reported in two studies, comprising a key risk factor for sphincteric injection [99], [100].

5.7.2 Benign Prostatic Hypertrophy/Prostatitis

Since surgical denervation has been shown to produce profound atrophy of the rat prostate and BoNT-A injection produces a long-term chemical denervation, it was logical that prostatic injection of BoNT-A could be used to potentially treat disorders associated with prostatic enlargement. Doggweiler et al. injected 30 rat prostates with BoNT-A and found that the total prostate volume (TPV) and weight were reduced, with a generalized glandular atrophy observed after histological staining [101]. Clinical outcomes in humans were reported soon thereafter for prostatitis and benign prostatic hypertrophy (BPH) [102], [103]. In a randomized, placebo-controlled study, Maria et al. enrolled 30 men with BPH to receive 4 mL saline injection or 200 U BoNT-A [103]. After 2 months, significantly more patients in the treated group had subjective symptomatic relief than in the control group. In patients who received BoNT-A, the symptom score was reduced by 65 % compared with baseline values and the serum prostate-specific antigen (PSA) concentration decreased by 51 % from baseline. In patients who received saline, the symptom score and PSA concentration were not significantly changed compared with the baseline values and 1-month values. No local complications or systemic side effects were observed in any patient.

Silva et al. investigated the duration of effect after a single 200 U BoNT-A injection into the prostate in 21 elderly men in urinary retention [104]. The authors found that TPV reached a nadir by 6 months and recovered to baseline by 18 months. Albeit nonsignificant, serum PSA showed a 25 % decrease from baseline to month 6. The 11 patients who were followed up for 18 months resumed spontaneous voiding at month 1. Mean maximal urinary flow rate (Q_{max}) increased while PSA and PVR decreased. In another study by Chuang et al., 41 men with a $Q_{max} < 12$ mL/sec or international prostate symptom score (IPSS) ≥ 8 underwent BoNT-A injection [105]. Injection dosages were 100 U (21 men, prostate volume < 30 mL) or 200 U (20 men, prostate volume > 30 mL) into the prostate transperineally under TRUS guidance. There were no significant local or systemic adverse effects in any man.

LUTS and QOL indices improved by $> 30\%$ in 76% of the men, and four of five men with urinary retention for over a month could void spontaneously at 1 week to 1 month after the BoNT-A injection. In 29% of the men there was no change in prostate volume, yet seven of these men still had a $> 30\%$ improvement in Q_{max}, LUTS and QOL. The efficacy was sustained at 12 months. Kuo suggested the use of BoNT-A injection in men with BPH who were poor surgical candidates [106]. All ten of his patients had an improvement in spontaneous voiding after treatment. Eight had an excellent result and two were improved. Both voiding pressure and PVR were significantly decreased after treatment. TPV was significantly reduced, and Q_{max} was significantly increased after treatment. The maximal effects of BoNT-A appeared at about 1 week and were maintained at 3 and 6 months after treatment. At 6–12 months of follow-up, no patient had recurrence of urinary retention, and the voiding condition in all patients remained at the posttreatment status.

The role of BoNT-A has been investigated as an add-on in men with TPV > 60 mL and an unsatisfactory response to combined alpha-blocker and 5-alpha-reductase inhibitor therapy [107]. Thirty men were randomized to BoNT-A with combination therapy, while 30 controls remained on combination therapy. Significant decreases in IPSS, Quality of Life Index (QOL-I) and TPV, and increase in Q_{max} were observed at 6 months and remained stable at 12 months in the treatment group. Improvements in IPSS and QOL-I were also observed at 6 months and a decrease in TPV at 12 months was noted in the control group. However, no significant changes in any parameters except for QOL-I at 6 and 12 months were noted between the treatment and control groups. AUR developed in three patients receiving BoNT-A treatment. Three BoNT-A and two medical treatment patients converted to transurethral surgery at the end of the study. Thus, the therapeutic effect of BoNT-A add-on was similar to combination medical treatment at 12 months. Chuang et al. also demonstrated that BoNT-A injection may provide therapeutic benefit in men with small prostates and LUTS [108]. In 16 men with a TPV < 30 mL and $Q_{max} < 12$ mL/sec, subjective improvement was seen at 1 week and maximal effect was achieved after 1 month. Benefits, such as improvement in MPV, IPSS, QOL-I and Q_{max}, were maintained at 3 and 6 months of follow-up. In two men who underwent prostate biopsy 1 month following BoNT-A injection, special staining demonstrated an increase in apoptotic activity not only in the glandular component but also in the stromal component of the prostatic tissue.

Oeconomou et al. performed a literature review consisting of the MEDLINE database as well as abstracts from several international conferences for studies regarding the use of BoNT-A for BPH [109]. Five experimental studies and ten clinical studies were found. The level of evidence was 1b for one study and 3 for the other studies, with grades of recommendation of A and C, respectively. The experimental studies reported induced relaxation of the prostate, atrophy and reduction of its size through inhibition of the trophic effect of the autonomic system on the prostate gland. In the clinical studies, all patients had LUTS due to benign prostatic enlargement, and TPV varied from < 20 mL to > 80 mL. The injection dose varied from 100 to 300 U and injection was performed transperineally, transrectally or transurethrally under general, local or without anesthesia. The follow-up period was 3–19.8 months.

All studies reported an improvement of Q_{max}, QOL-I, and reduction of IPSS, PSA, PVR and TPV. Local or systemic side effects were rare. Only patients with retention needed a urethral drainage catheter. Additional studies published since this review have described similar outcomes [110].

Finally, Zermann et al. described a benefit from perisphincteric injection of 200 U of BoNT-A in 11 men with chronic prostatic pain [111]. All of the patients suffered from a pathological pelvic floor tenderness, an inability of sufficient conscious pelvic floor control, a urethral hypersensitivity/hyperalgesia and a urethral muscle hyperactivity. Basic parameters of bladder function (capacity, sensitivity, compliance) were normal. The BoNT-A injection was followed by a pelvic floor muscle weakening and a relief of prostatic pain and urethral hypersensitivity/hyperalgesia.

In conclusion, the injection of BoNT-A into the prostate represents an alternative and minimally invasive treatment for LUTS associated with BPH. Intraprostatic BoNT-A injection induced prostate degeneration and apoptosis in the animal model, and, in clinical studies, BoNT-A injection improves Q_{max}, TPV, PSA, PVR, IPSS and QOL indices. This therapy may be a promising alternative for medicine-refractory BPH, poor surgical candidates and prostatitis. Local and systemic side effects are uniformly rare. It should be recognized that this therapy remains experimental and, while encouraging, the overall level of clinical evidence at this time is low to support the routine use of BoNT-A for intraprostatic injection.

5.7.3   *Interstitial Cystitis/Pelvic Pain*

The therapeutic benefits of BoNT-A injection in the lower urinary tract may not be limited to muscular relaxation and may extend to pain disorders. While the exact mechanism by which BoNT-A exerts its analgesic effects in humans is incompletely understood, it has been shown that BoNT-A inhibits the release of substance P from rat dorsal root ganglia [112]. BoNT-A was the most potent of all of the BoNTs tested and significant inhibition of substance P release was observed after 4 h of incubation with continued increase for 8 h. Inhibition then stabilized and remained at steady levels for 15 days. Chuang et al. also found that intravesical BoNT-A instillation in a rat model blocked acetic acid-induced bladder pain responses and inhibited CGRP release from afferent nerve terminals [113]. Likewise, BoNT-A administration has been found to modulate other mediators of the inflammatory response, such as ATP and cyclooxygenase-2, further providing evidence of possible effectiveness in painful bladder conditions [114], [115].

Several studies have been conducted implementing BoNT-A in interstitial cystitis (IC)/painful bladder syndrome (PBS). Smith et al. treated 13 women with IC (as defined by the criteria of the National Institute of Diabetes, Digestive and Kidney Disease) with 100–200 U in 20–30 sites [116]. The patients injected in Poland had Dysport® and the trigone was included in the injection. Overall, 9 (69 %) of 13 patients noted subjective improvement after BoNT-A treatment. The Interstitial Cystitis Symptom Index and Interstitial Cystitis Problem Index mean scores significantly improved by 71 and 69 %, respectively. Daytime frequency, nocturia and pain by visual analog scale decreased significantly by 44, 45 and 79 %, respectively. The first desire to void and MCC in the seven Polish patients significantly increased by 58 and 57 %, respectively. Giannantoni et al. reported subjective improvement in 13 of 15 patients injected with 200 U BoNT-A for IC at 1-month and 3-month follow-up [117]. The mean visual analog score (VAS), and daytime and nighttime urinary frequency (UF) were all significantly decreased. At the 5-month follow-up, the beneficial effects persisted in 26.6 % of cases, and at 12 months, pain recurred at baseline levels in all patients. Nine patients complained of dysuria 1 month after treatment and dysuria persisted in four patients at the 3-month follow-up and in two at the 5-month follow-up. Two patients developed AUR and required short-term CIC. Kuo and Chancellor compared the clinical effectiveness of BoNT-A injections followed by hydrodistention (HD) with HD alone in patients with IC/PBS [118]. This was a prospective, randomized study that enrolled 67 patients with IC/PBS who had failed conventional treatments. Of these, 44 patients received suburothelial injection with 100–200 U BoNT-A followed by cystoscopic HD 2 weeks later, while 23 patients (control group) received the HD procedure only. The IC/PBS symptom score significantly decreased in all groups, but VAS reduction, functional bladder capacity (FBC) and MCC increases were significant only in the BoNT-A group at 3 months. Of the 44 patients in the BoNT-A group, 31 (71 %) had a successful result at 6 months. A successful result at 12 and 24 months was reported in 24 (55 %) and 13 (30 %) patients in BoNT-A group, respectively, compared with only six (26 %) and four (17 %) in the control group.

BoNT-A injection may also benefit patients with radiation and chemical cystitis [119]. Four men with refractory radiation cystitis after prostate radiation (two underwent external beam radiation and two underwent brachytherapy) and two women with external beam radiation for cervical cancer were treated with 200 U BoNT-A detrusor injection. Two patients with refractory Bacillus Calmette–Guerin (BCG) cystitis were also treated with 100 U bladder BoNT-A injections. Of the six patients treated for radiation cystitis, one patient noted no improvement, three reported moderate improvement and two had significant improvement in their symptoms. Mean bladder capacity (MBC) more than doubled to 250 mL, urinary frequency decreased and PVR remained similar compared to baseline. Similar outcomes were observed in the two patients with BCG cystitis, including a decrease in the VAS for pain from eight to two.

5.7.4   Urethral Stricture Disease

Owing to the success of BoNT-A in preventing scar formation in facial wounds, this product was attempted to decrease the scar formation in men with recurrent posterior urethral strictures [120], [121]. Three men with variable etiologies for their stricture disease underwent repeat direct visual internal urethrotomy (DVIU) and BoNT-A injection. Using a 25-gauge Williams needle, 100 U BoNT-A diluted in 2 mL saline was injected circumferentially at the base of the scar into the intervening areas between DVIU incisions. Two of the three men had no recurrence of stricture and subsequently underwent uneventful artificial urethral sphincter implantation. The third man had a moderate recurrence of his stricture and the lumen diameter stabilized at 10–12F by 6 weeks. At 9 months, he was voiding to completion without the need for catheterization.

5.7.5 Detrusor Underactivity

Kuo investigated the effects of BoNT-A in treating patients with voiding dysfunction and concomitant detrusor underactivity [122]. Twenty patients with chronic urinary retention or severe dysuria received 50 U of BoNT-A via urethral injection into the external sphincter. Among the 90 % of patients with satisfactory results, mean QOL score improved significantly, as did urodynamic indices such as mean voiding pressure (MVP), maximal urethral closure pressure and PVR. The subjective maximal improvement was observed by 1–2 weeks after treatment and was maintained for 3 months. In seven patients, indwelling catheters were removed, and in the four patients performing CIC, the frequency decreased or CIC was discontinued. The remaining seven patients with voiding difficulty had a significant improvement in the obstructive symptom score.

5.7.6 Pediatric NDO

Schulte-Baukloh et al. initially demonstrated encouraging preliminary outcomes in children with neurogenic bladder refractory to antimuscarinic medications [123], [124]. A weight-adapted dose of 85–300 U BoNT-A was injected into 40–50 sites in the detrusor. All 20 children underwent follow-up cystometry within 4 weeks of injection with improvement in urodynamic indices. Specifically, the MBP decreased by 32.6 %, BRV more than doubled and MCC increased by 56 % in comparison to baseline values. Similar positive improvements in urodynamic parameters have since been reported in additional studies of pediatric neurogenic bladder after undergoing BoNT-A injection [125], [126]. A pediatric patient with bladder dysfunction due to posterior urethral valves improved to a similar degree as patients with NDO [126].

Hoebeke et al. evaluated the outcomes of BoNT-A injection in 21 children with therapy-resistant non-neurogenic DO [127]. One girl had temporary urinary retention (< 10 days) and one boy had signs of VUR with flank pain during voiding, which spontaneously resolved after 2 weeks. Two girls experienced one episode each of symptomatic lower UTI. Eight girls and seven boys with a minimum follow-up of 6 months represented the study group for long-term evaluation. After one injection, nine patients showed full response (resolution of urge and daytime dryness) with a significant mean increase in BC. Three patients had a partial response (50 % decrease in urgency and incontinence) and three remained unchanged. Eight of the nine full responders were still cured after 12 months, while one of the initially successfully treated patients had a relapse after 8 months. The three partial responders and the one relapser underwent a second injection, with a full response in the former full responder and in one partial responder.

Several groups have shown a benefit from repeated BoNT-A injections in children. Schulte-Baukloh et al. evaluated the long-term outcomes with repeated BoNT-A injection in ten children with NDO [128]. All received at least three detrusor injections and four had received five or more. The relative changes in urodynamic parameters obtained 6 months after each injection, in comparison with preinjection values, after the first versus fifth injection were as follows: increase in BRV (81 vs. 88 %), decrease in MDP (7 vs. 39 %) and increase in MCC (88 vs. 72 %). Bladder compliance showed no change 6 months after the first injection and increased by 109 % after the fifth injection. No major treatment-related adverse events were reported. At a mean follow-up of 41 months, Romero et al. found that BC, detrusor accommodation and pressure improved after 4 weeks in all but two of 12 patients [126]. This improvement decreased after 6 months, although successive injections produced similar changes. Of 12 patients, 11 received a total of two or three injections, with clinical and urodynamic improvement in eight patients preventing bladder augmentation. Altaweel et al. also evaluated the effect of repeated BoNT-A injections in 20 children with NDO due to myelomeningocele unresponsive to medical management [129]. Of the 20, 13 became continent and MBC increased significantly while MDP decreased significantly. Compliance also increased. At a mean of 8 months, all 13 patients underwent a second injection which led to similar improvements in MBC, MDP and compliance. Of the seven patients failing to improve after first injection, six failed to improve after a second injection. Schulte-Baukloh et al. suggested that patients who show a failure of therapy after BoNT-A injections for which no other causes can be determined should have their serum checked for BoNT-A antibodies [130]. Recurrent UTI might be a predisposing factor for BoNT/A antibodies. Finally, a retrospective study of seven children with NDO who received 18 injections revealed that social continence was achieved after the first injection and maximum catheterized volume and BRV increased while MDP decreased [131].

A recent study attempted to evaluate whether electromotive administration of BoNT-A was feasible in children [132]. During instillation of intravesical BoNT-A, a pulsed current generator delivered 10 mA for 15 min through a specially designed catheter. Significant improvement was observed in MCC, BRV, P_{detmax} and end-filling pressure. Urinary incontinence was improved in 12 of 15 children and none required surgical intervention. Skin erythema and burning sensation were observed in six children.

5.8 Conclusions

A recently published European consensus report on the use of BoNT-A in the treatment of lower urinary tract disorders came to several conclusions [133]. Employing the European Association of Urology levels of evidence, the panel found grade A evidence to support the use of BoNT-A for intractable symptoms of IDO or NDO in adults. The panel recommended caution in patients with IDO, as the risk of voiding difficulty and duration of effect have not yet been accurately evaluated. There was grade B evidence to support repeated treatment in patients with NDO. The depth and location for bladder injections should be within the detrusor and injections should spare the trigone (grade C). Dosage in children should be determined by body weight, with caution regarding total dose if also being used for the treatment of spasticity, and minimum age (grade B). Existing evidence was inconclusive for recommendations in neurogenic DSD, bladder pain syndrome, prostate diseases and pelvic floor disorders. The panel concluded that the use of BoNT-A in the lower urinary tract with the current dosages and techniques was considered to be safe overall (grade A). As mentioned previously, BoNT-A has been approved for both IDO and NDO in adults by the US FDA.

It is clear that, while not all of the mechanisms of action are completely understood in humans, BoNT-A is an appealing option for many disorders of the lower urinary tract. The adverse event profile is favorable, with severe adverse events (AEs) occurring infrequently. Improvements in bladder diary variables, urodynamic indices and QOL indicators are uniformly seen. It must be noted that the effects of BoNT-A injection wane with time and reinjections are invariably necessary to maintain benefits.

References

1. Van Ermengem E (1897) Ueber einen neunen anaeroben Bacillus and seine Beziehungen zum Botulisms. Ztsch Hyg Infect 26:1
2. Gill DM (1982) Bacterial toxins: a table of lethal amounts. Microbiol Rev 46:86
3. Smith CP, Chancellor MB (2004) Emerging role of botulinum toxin in the treatment of voiding dysfunction. J Urol 171:2128–2137
4. Shapiro RL, Hatheway C, Swerdlow DL (1998) Botulism in the United States: a clinical and epidemiologic review. Ann Intern Med 129:221
5. Naumann M, So Y, Argoff CE, Childers MK, Dykstra DD, Gronseth GS et al (2008) Assessment: Botulinum neurotoxin in the treatment of autonomic disorders and pain (an evidence-based review): report of the Therapeutics and Technology Assessment Subcommittee of the American Academy of Neurology. Neurology 70:1707–1714
6. DasGupta BRD (1994) Structures of botulinum neurotoxin, its functional domains, and perspectives on the crystalline type A toxin. In: Janovich J, Hallet M (eds) Therapy with botulinum toxin. Marcel Dekker, New York
7. Schiavo G, Papini E, Genna G et al (1990) An intact interchain disulfide bond is required for the neurotoxicity of tetanus toxin. Infect Immun 58:4136–4141
8. de Paiva A, Ashton AC, Foran P et al (1993) Botulinum A like type B and tetanus toxins fulfill criteria for being a zinc dependent protease. J Neurochem 61:2338–2341
9. Brin MF (1997) Botulinum toxin: chemistry, pharmacology, toxicity, and immunology. Muscle Nerve 6:S146
10. Oguma K, Fujinaga Y, Inoue K (1995) Structure and function of Clostridium botulinum toxins. Microbiol Immunol 39:161
11. Montecucco C, Schiavo G, Tugnoli V et al (1996) Botulinum neurotoxins: mechanism of action and therapeutic applications. Mol Med Today 2:418
12. Rothman J (1994) Mechanism of intracellular protein transport. Nature 372:55

13. Rosetto O, Deloye F, Poulain B et al (1995) The metallo-proteinase activity of tetanus and botulism neurotoxins. J Physiol Paris 89:43

14. Schiavo G, Matteoli M, Montecucco C (2000) Neurotoxins affecting neuroexocytosis. Physiol Rev 80:717–766

15. de Paiva A, Meunier FA, Molgo J et al (1999) Functional repair of motor endplates after botulinum neurotoxin type A poisoning: biphasic switch of synaptic activity between nerve sprouts and their parent terminals. Proc Natl Acad Sci USA 96:3200–3205

16. Angaut-Petit D, Molgo J, Comella JX et al (1990) Terminal sprouting in mouse neuromuscular junctions poisoned with botulinum type A toxin: morphological and electrophysiological features. Neuroscience 37:799–808

17. Sahai A, Khan M, Fowler CJ, Dasgupta P (2005) Botulinum toxin for the treatment of lower urinary tract symptoms. Neurourol Urodynam 24:2–12

18. Meunier F, Schiavo G, Molgo J (2002) Botulinum neurotoxins: from paralysis to recovery of functional neuromuscular transmission. J Physiol Paris 96:105–113

19. Haferkamp A, Schurch B, Reitz A et al (2004) Lack of ultrastructural detrusor changes following endoscopic injection of botulinum toxin type A in overactive neurogenic bladder. Eur Urol 46:784–791

20. Naumann M, Jost WH, Toyka KV (1999) Botulinum toxin in the treatment of neurological disorders of the autonomic nervous system. Arch Neurol 56:914–916

21. Duggan MJ, Quinn CP, Chaddock JA et al (2002) Inhibition of release of neurotransmitters from rat dorsal root ganglia by a novel conjugate of a Clostridium botulinum toxin A endopeptidase fragment and Erythrina cristagalli lectin. J Biol Chem 277:34846–34852

22. Rapp DE, Turk KW, Bales GT et al (2006) Botulinum toxin type a inhibits calcitonin gene-related peptide release from isolated rat bladder. J Urol 175:1138–1142

23. Apostolidis A, Popat R, Yiangou Y et al (2005) Decreased sensory receptors P2X3 and TRPV1 in suburothelial nerve fibers following intradetrusor injections of botulinum toxin for human detrusor overactivity. J Urol 174:977–982

24. Simpson LL (2004) Identification of the major steps in botulinum toxin action. Annu Rev Pharmacol Toxicol 44:167–193

25. Mouttalib S, Khan S, Castel-Lacanal E, Guillotreau J, De Boissezon, Malavaud B et al (2010) Risk of urinary tract infection after detrusor botulinum toxin A injections for refractory neurogenic detrusor overactivity in patients with no antibiotic treatment. BJU Int 106:1677–1680

26. Schurch B, Reitz A, Tenti G (2004) Electromotive drug administration of lidocaine to anesthetize the bladder before botulinum-A toxin injections into the detrusor. Spinal Cord 42:338–341

27. Chuang YC, Chancellor MB (2006) The application of botulinum toxin in the prostate. J Urol 176:2375–2382

28. Chuang YC, Chiang PH, Huang CC, Yoshimura N, Chancellor MB (2005) Botulinum toxin type A improves benign prostatic hyperplasia symptoms in patients with small prostates. Urology 66:775

29. Chuang YC, Tu CH, Huang CC, Lin HJ, Chiang PH, Yoshimura N et al (2006) Intraprostatic injection of botulinum toxin type A injection relieves bladder outlet obstruction and induces prostate apoptosis. BMC Urol 6:12

30. Radziszewski P, Dobronski P, Borkowski A (2001) Treatment of the non-neurogenic storage and voiding disorders with the chemical denervation caused by botulinum toxin type A—a pilot study. Neurourol Urodynam 20:410–412

31. Dmochowski R, Chapple C, Nitti VW, Chancellor M, Everaert K, Thompson C, Daniell G, Zhou J, Haag- Molkenteller C (2010) Efficacy and safety of onabotulinumtoxinA for idiopathic overactive bladder: a double- blind, placebo controlled, randomized dose ranging trial. J Urol 184:2416–2422

32. Rovner E, Kennelly M, Schulte-Baukloh H, Zhou J, Haag-Molkenteller C, Dasgupta P (2011) Urodynamic results and clinical outcomes with intradetrusor injections of onabotulinumtoxina

in a randomized, placebo-controlled, dose-finding study in idiopathic overactive bladder. Neurourol Urodynam 30:556–562

33. Tincello DG, Kenyon S, Abrams KR, Mayne C, Toozs-Hobson P, Taylor D et al (2012) Botulinum toxin a versus placebo for refractory detrusor overactivity in women: a randomised blinded placebo-controlled trial of 240 women (the RELAX study). Eur Urol 62:507–514

34. Nitti VW, Dmochowski R, Herschorn S, Sand P, Thompson C, Nardo C et al, EMBARK Study Group (2013) OnabotulinumtoxinA for the treatment of patients with overactive bladder and urinary incontinence: results of a phase 3, randomized, placebo controlled trial. J Urol 189:2186–2193

35. Chapple C, Sievert KD, Macdiarmid S, Khullar V, Radziszewski P, Nardo C et al (2013) OnabotulinumtoxinA 100 U significantly improves all idiopathic overactive bladder symptoms and quality of life in patients with overactive bladder and urinary incontinence: a randomised, double-blind, placebo-controlled trial. Eur Urol 2013 64:249–256

36. Fowler CJ, Auerbach S, Ginsberg D, Hale D, Radziszewski P, Rechberger T et al (2012) OnabotulinumtoxinA improves health-related quality of life in patients with urinary incontinence due to idiopathic overactive bladder: a 36-week, double-blind, placebo-controlled, randomized, dose-ranging trial. Eur Urol 62:148–157

37. Brubaker L, Gousse A, Sand P, Thompson C, Patel V, Zhou J et al (2012) Treatment satisfaction and goal attainment with onabotulinumtoxinA in patients with incontinence due to idiopathic OAB. Int Urogynecol J 23:1017–1025

38. Schmid DM, Sauermann P, Werner M, Schuessler B, Blick N, Muentener M et al (2006) Experience with 100 cases treated with botulinum-A toxin injections in the detrusor muscle for idiopathic overactive bladder syndrome refractory to anticholinergics. J Urol 176:177–185

39. Sahai A, Khan MS, Le Gall N, Dasgupta P, on behalf of the GKT Botulinum Study Group (2008) Urodynamic assessment of poor responders after botulinum toxin-A treatment for overactive bladder. Urology 71:455–459

40. Makovey I, Davis T, Guralnick ML, O'Connor RC (2011) Botulinum toxin outcomes for idiopathic overactive bladder stratified by indication: lack of anticholinergic efficacy versus intolerability. Neurourol Urodyn 30:1538–1540

41. Khan S, Panicker J, Roosen A, Gonzales G, Elneil S, Dasgupta P, Fowler CJ, Kessler TM (2010) Complete continence after botulinum neurotoxin type A injections for refractory idiopathic detrusor overactivity incontinence: patient-reported outcome at 4 weeks. Eur Urol 57:891–896

42. Sahai A, Dowson C, Khan MS, Dasgupta P, GKT Botulinum Study Group (2010) Repeated injections of botulinum toxin-A for idiopathic detrusor overactivity. Urology 75:552–558

43. Khan S, Kessler TM, Apostolidis A, Kalsi V, Panicker J, Roosen A, Gonzales G, Haslam C et al (2009) What a patient with refractory idiopathic detrusor overactivity should know about botulinum neurotoxin type A injections. J Urol 181:1773–1778

44. Game X, Khan S, Panicker JN, Kalsi V, Dalton C, Elneil S, Hamid R, Dasgupta P, Fowler CJ (2011) Comparison of the impact on health-related quality of life of repeated detrusor injections of botulinum toxin in patients with idiopathic or neurogenic detrusor overactivity. BJU Int 107:1786–1792

45. Dowson C, Watkins J, Khan MS, Dasgupta P, Sahai A (2012) Repeated botulinum toxin type A injections for refractory overactive bladder: medium-term outcomes, safety profile, and discontinuation rates. Eur Urol 61:834–839

46. Granese R, Adile G, Gugliotta G, Cucinella G, Saitta S, Adile B (2012). Botox$^{®}$ for idiopathic overactive bladder: efficacy, duration and safety. Effectiveness of subsequent injection. Arch Gynecol Obstet 286:923–929

47. Dowson C, Khan MS, Dasgupta P, Sahai A (2010) Repeat botulinum toxin-A injections for treatment of adult detrusor overactivity. Nat Rev Urol 7:661–667

48. Gormley EA, Lightner DJ, Burgio KL, Chai TC, Clemens JQ, Culkin DJ et al (2012) Diagnosis and treatment of overactive bladder (non-neurogenic) in adults: AUA/SUFU guideline. J Urol 188:2455–2463

49. Schurch B, Stöhrer M, Kramer G, Schmid DM, Gaul G, Hauri D (2000) Botulinum-A toxin for treating detrusor hyperreflexia in spinal cord injured patients: a new alternative to anticholinergic drugs? Preliminary results. J Urol 164:692–697
50. Schurch B, Schmid DM, Stohrer M (2000) Treatment of neurogenic incontinence with botulinum toxin A. N Engl J Med 342:665
51. Schurch B, de Sèze M, Denys P, Chartier-Kastler E, Haab F, Everaert K et al (2005) Botox Detrusor Hyperreflexia Study Team. Botulinum toxin type a is a safe and effective treatment for neurogenic urinary incontinence: results of a single treatment, randomized, placebo controlled 6-month study. J Urol 174:196–200
52. Reitz A, Stohrer M, Kramer G, Del Popolo G, Chartier-Kastler E, Pannek J et al (2004) European experience of 200 cases treated with botulinum-A toxin injections into the detrusor muscle for urinary incontinence due to neurogenic detrusor overactivity. Eur Urol 45:510–515
53. Bagi P, Biering-Sorensen F (2004) Botulinum toxin A for treatment of neurogenic detrusor overactivity and incontinence in patients with spinal cord lesions. Scand J Urol Nephrol 38:495–498
54. Kessler TM, Danuser H, Schumacher M, Studer UE, Burkhard FC (2005) Botulinum A toxin injections into the detrusor: an effective treatment in idiopathic and neurogenic detrusor overactivity? Neurourol Urodyn 24:231–236
55. Popat R, Apostolidis A, Kalsi V, Gonzales G, Fowler CJ, Dasgupta P (2005) A comparison between the response of patients with idiopathic detrusor overactivity and neurogenic detrusor overactivity to the first intradetrusor injection of botulinum-A toxin. J Urol 174:984–989
56. Hajebrahimi S, Altaweel W, Cadoret J, Cohen E, Corcos J (2005) Efficacy of botulinum-A toxin in adults with neurogenic overactive bladder: initial results. Can J Urol 12:2543–2546
57. Klaphajone J, Kitisomprayoonkul W, Sriplakit S (2005) Botulinum toxin type A injections for treating neurogenic detrusor overactivity combined with low-compliance bladder in patients with spinal cord lesions. Arch Phys Med Rehabil 86:2114–2118
58. Kuo HC (2006) Therapeutic effects of suburothelial injection of botulinum a toxin for neurogenic detrusor overactivity due to chronic cerebrovascular accident and spinal cord lesions. Urology 67:232–236
59. Schulte-Baukloh H, Schobert J, Stolze T, Sturzebecher B, Weiss C, Knispel HH (2006) Efficacy of botulinum-A toxin bladder injections for the treatment of neurogenic detrusor overactivity in multiple sclerosis patients: an objective and subjective analysis. Neurourol Urodynam 25:110–115
60. Tow AM, Toh KL, Chan SP, Consigliere D (2007) Botulinum toxin type A for refractory neurogenic detrusor overactivity in spinal cord injured patients in Singapore. Ann Acad Med Singapore 36:11–17
61. Stoehrer M, Wolff A, Kramer G, Steiner R, Lmochner-Ernst D, Leuth D et al (2009) Treatment of neurogenic detrusor overactivity with botulinum toxin A: the first seven years. Urol Int 83:379–385
62. Giannantoni A, Rossi A, Mearini E, Del Zingaro M, Porena M, Berardelli A (2009) Botulinum toxin A for overactive bladder and detrusor muscle overactivity in patients with Parkinson's disease and multiple system atrophy. J Urol 182:1453–1457
63. Alvares RA, Silva JA, Barboza AL, Monteiro RT (2010) Botulinum toxin A in the treatment of spinal cord injury patients with refractory neurogenic detrusor overactivity. Int Braz J Urol 36:732–737
64. Giannantoni A, Mearini E, Del Zingaro M, Porena M (2009) Six-year follow-up of botulinum toxin A intradetrusorial injections in patients with refractory neurogenic detrusor overactivity: clinical and urodynamic results. Eur Urol 55:705–711
65. Wefer B, Ehlken B, Bremer J, Burgdorfer H, Domurath B, Hampel C et al (2010) Treatment outcomes and resource use of patients with neurogenic detrusor overactivity receiving botulinum toxin A (BOTOX) therapy in Germany. World J Urol 28:385–390
66. Mehnert U, Birzele J, Reuter K, Schurch B (2010) The effect of botulinum toxin type A on overactive bladder symptoms in patients with multiple sclerosis. J Urol 184:1011–1016

67. Gomes CM, de Castro Filho JE, Rejowski RF, Trigo-Rocha FE, Bruschini H, de Barros Filho TE, Srougi M (2010) Experience with different botulinum toxins for the treatment of refractory neurogenic detrusor overactivity. Int Braz J Urol 36:66–74

68. Chen G, Liao L (2011) Injections of botulinum toxin A into the detrusor to treat neurogenic detrusor overactivity secondary to spinal cord injury. Int Urol Nephrol 43:655–662

69. Chen CY, Liao CH, Kuo HC (2011) Therapeutic effects of botulinum toxin A injection on neurogenic detrusor overactivity in patients with different levels of spinal cord injury and types of detrusor sphincter dyssynergia. Spinal Cord 49:659–664

70. Herschorn S, Gajewski J, Ethans K, Corcos J, Carlson K, Bailly G et al (2011) Efficacy of botulinum toxin A injection for neurogenic detrusor overactivity and urinary incontinence: a randomized, double-blind trial. J Urol 185:2229–2235

71. Cruz F, Herschorn S, Aliotta P, Brin M, Thompson C, Lam W et al (2011) Efficacy and safety of onabotulinumtoxinA in patients with urinary incontinence due to neurogenic detrusor overactivity: a randomised, double-blind, placebo-controlled trial. Eur Urol 60:742–750

72. Sussman D, Patel V, Del Popolo G, Lam W, Globe D, Pommerville P (2013) Treatment satisfaction and improvement in health-related quality of life with onabotulinumtoxinA in patients with urinary incontinence due to neurogenic detrusor overactivity. Neurourol Urodyn 32:242–249

73. Kennelly M, Dmochowski R, Ethans K, Karsenty G, Schulte-Baukloh H, Jenkins B, Thompson C et al (2013) Long-term efficacy and safety of onabotulinumtoxinA in patients with urinary incontinence due to neurogenic detrusor overactivity: an interim analysis. Urology 81:491–497

74. Hori S, Patki P, Attar KH, Ismail S, Vasconcelos JC, Shah PJ (2009) Patients' perspective of botulinum toxin-A as a long-term treatment option for neurogenic detrusor overactivity secondary to spinal cord injury. BJU Int 104:216–220

75. Kalsi V, Apostolidis A, Popat R, Gonzales G, Fowler CJ, Dasgupta P (2006) Quality of life changes in patients with neurogenic versus idiopathic detrusor overactivity after intradetrusor injections of botulinum neurotoxin type A and correlations with lower urinary tract symptoms and urodynamic changes. Eur Urol 49:528–535

76. Schurch B, Denys P, Kozma CM, Reese PR, Slaton T, Barron RL (2007) Botulinum toxin A improves the quality of life of patients with neurogenic urinary incontinence. Eur Urol 52:850–858

77. Karsenty G, Denys P, Amarenco G, De Seze M, Game X, Haab F et al (2008) Botulinum toxin A (Botox) intradetrusor injections in adults with neurogenic detrusor overactivity/neurogenic overactive bladder: a systematic literature review. Eur Urol 53:275–287

78. Karsenty G, Reitz A, Lindemann G, Boy S, Schurch B (2006) Persistence of therapeutic effect after repeated injections of botulinum toxin type A to treat incontinence due to neurogenic detrusor overactivity. Urology 68:1193–1197

79. Chenet A, Perrouin-verbe B, Le Normand L, Labat JJ, Brunel P, Lefort M et al (2007) Efficacy of repeat injections of botulinum A toxin to the detrusor in neurogenic bladder overactivity. Ann Readapt Med Phys 50:651–660

80. Reitz A, Denys P, Fermanian C, Schurch B, Comperat E, Chartier-Kastler E (2007) Do repeat intradetrusor botulinum toxin type a injections yield valuable results? Clinical and urodynamic results after five injections in patients with neurogenic detrusor overactivity. Eur Urol 52:1729–1735

81. Pannek J, Gocking K, Bersch U (2009) Long-term effects of repeated intradetrusor botulinum neurotoxin A injections on detrusor function in patients with neurogenic bladder dysfunction. BJU Int 104:1246–1250

82. Karsenty G, Elzayat E, Delapparent T, St-Denis B, Lemieux MC, Corcos J (2007) Botulinum toxin type a injections into the trigone to treat idiopathic overactive bladder do not induce vesicoureteral reflux. J Urol 177:1011–1014

83. Mascarenhas F, Cocuzza M, Gomes CM, Leao N (2008) Trigonal injection of botulinum toxin-A does not cause vesicoureteral reflux in neurogenic patients. Neurourol Urodynam 27:311–314

84. Abdel-Meguid TA (2010) Botulinum toxin-A injections into neurogenic overactive bladder—to include or exclude the trigone? A prospective, randomized, controlled trial. J Urol 184:2423–2428

85. Game X, Castel-Lacanal E, Bentaleb Y, Thiry-Escudie I, De Boissezon X, Malavaud B et al (2008) Botulinum toxin A detrusor injections in patients with neurogenic detrusor overactivity significantly decrease the incidence of symptomatic urinary tract injections. Eur Urol 53:613–618

86. Nitti VW (2006) Botulinum toxin for the treatment of idiopathic and neurogenic overactive bladder: state of the art. Rev Urol 8:198–208

87. Kuo HC, Liao CH, Chung SD (2010) Adverse events of intravesical botulinum toxin A injections for idiopathic detrusor overactivity: risk factors and influence on treatment outcome. Eur Urol 58:919–926

88. Brubaker L, Richter HE, Visco A, Mahajan S, Nygaard I, Braun TM, Barber MD et al (2008) Refractory idiopathic urge urinary incontinence and botulinum A injection. J Urol 180:217–222

89. Kessler TM, Khan S, Panicker J, Roosen A, Elneil S, Fowler CJ (2009) Clean intermittent self-catheterization after botulinum neurotoxin type A injections: short-term effect on quality of life. Obstet Gynecol 113:1046–1051

90. Wyndaele JJ, Van Dromme SA (2002) Muscular weakness as side effect of botulinum toxin injection for neurogenic detrusor overactivity. Spinal Cord 40:599–600

91. Bauer RM, Gratzke C, Roosen A, Hocaoglu Y, Mayer ME, Buchner A et al (2011) Patient-reported side effects of intradetrusor botulinum toxin type A for idiopathic overactive bladder syndrome. Urol Int 86:68–72

92. Naumann M, Jankovic J (2004) Safety of botulinum toxin type A: a systematic review and meta-analysis. Curr Med Res Opin 20:981–990

93. Dykstra DD, Sidi AA, Scott AB, Pagel JM, Goldish GD (1988) Effects of botulinum A toxin on detrusor-sphincter dyssynergia in spinal cord injury patients. J Urol 139:919–922

94. Schurch B, Hauri D, Rodic B, Curt A, Meyer M, Rossier AB (1996) Botulinum-A toxin as a treatment of detrusor-sphincter dyssynergia: a prospective study in 24 spinal cord injury patients. J Urol 155:1023–1029

95. Gallien P, Reymann JM, Amarenco G, Nicolas B, de Seze, M, Bellisant E (2005) Placebo controlled, randomised, double blind study of the effects of botulinum A toxin on detrusor sphincter dyssynergia in multiple sclerosis patients. J Neurol Neurosurg Psychiatry 76:1670–1676

96. Dykstra DD, Sidi AA (1990) Treatment of detrusor-sphincter dyssynergia with botulinum A toxin: a double-blind study. Arch Phys Med Rehabil 71:24–26

97. de Seze M, Petit H, Gallien P, de Seze MP, Joseph PA, Mazaux JM et al (2002) Botulinum a toxin and detrusor sphincter dyssynergia: a double-blind lidocaine-controlled study in 13 patients with spinal cord disease. Eur Urol 42:56–62

98. Liao YM, Kuo HC (2007) Causes of failed urethral botulinum toxin A treatment for emptying failure. Urology 70:763–766

99. Petit H, Wiart L, Gaujard E, LeBreton F, Ferriere JM, Lagueny A et al (1998) Botulinum A toxin treatment for detrusor-sphincter dyssynergia in spinal cord disease. Spinal Cord 36:91–94

100. Fowler CJ, Betts CD, Christmas TJ, Swash M, Fowler CG (1992) Botulinum toxin in the treatment of chronic urinary retention in women. Br J Urol 70:387–389

101. Doggweiler R, Zermann DH, Ishigooka M, Schmidt RA (1998) Botox-induced prostatic involution. Prostate 37:44–50

102. Maria G, Destito A, Lacquaniti S, Bentivoglio AR, Brisinda G, Albanese A (1998) Relief by botulinum toxin of voiding dysfunction due to prostatitis. Lancet 352:625

103. Maria G, Brisinda G, Civello IM, Bentivoglio AR, Sganga G, Albanese A (2003) Relief by botulinum toxin of voiding dysfunction due to benign prostatic hyperplasia: results of a randomized, placebo-controlled study. Urology 62:259–264

104. Silva J, Pinto R, Carvalho T, Botelho F, Silva P, Oliveira R et al (2009) Intraprostatic Botulinum Toxin Type A injection in patients with benign prostatic enlargement: duration of the effect of a single treatment. MBC Urol 9:9
105. Chuang YC, Chiang PH, Yoshimura N, De Miguel F, Chancellor MB (2006) Sustained beneficial effects of intraprostatic botulinum toxin type A on lower urinary tract symptoms and quality of life in men with benign prostatic hyperplasia. BJU Int 98:1033–1037
106. Kuo HC (2005) Prostate botulinum A toxin injection—an alternative treatment for benign prostatic obstruction in poor surgical candidates. Urology 65:670–674
107. Kuo HC, Liu HT (2009) Therapeutic effects of add-on botulinum toxin A on patients with large benign prostatic hyperplasia and unsatisfactory response to combined medical therapy. Scand J Urol Nephrol 43:201–211
108. Chuang YC, Chiang PH, Huang CC, Yoshimura N, Chancellor MB (2005) Botulinum toxin type A improves benign prostatic hyperplasia symptoms in patients with small prostates. Urology 66:775–779
109. Oeconomou A, Madersbacher H, Kiss G, Berger TJ, Melekos M, Rehder P (2008) Is botulinum neurotoxin type A (BoNT-A) a novel therapy for lower urinary tract symptoms due to benign prostatic enlargement? A review of the literature. Eur Urol 54:765–775
110. Brisinda G, Caddedu F, Vanella S, Mazzeo P, Marniga G, Maria G (2009) Relief by botulinum toxin of lower urinary tract symptoms owing to benign prostatic hyperplasia: early and long-term results. Urology 73:90–94
111. Zermann D, Ishigooka M, Schubert J, Schmidt RA (2000) Perisphincteric injection of botulinum toxin type A. A treatment option for patients with chronic prostatic pain? Eur Urol 38:393–399
112. Welch MJ, Purkiss JR, Foster KA (2000) Sensitivity of rat dorsal root ganglia neurons to clostridium botulinum neurotoxins. Toxicon 38:245–258
113. Chuang YC, Yoshimura N, Huang CC, Chiang PH, Chancellor MB (2004) Intravesical botulinum toxin a administration produces analgesia against acetic acid induced bladder pain responses in rats. J Urol 172:1529–1532
114. Chuang YC, Yoshimura N, Huang CC et al (2008) Intraprostatic botulinum toxin a injection inhibits cyclooxygenase-2 expression and suppresses prostatic pain on capsaicin induced prostatitis model in the rat. J Urol 180:742–748
115. Khera M, Somogyi GT, Kiss S et al (2004) Botulinum toxin A inhibits ATP release from bladder urothelium after chronic spinal cord injury. Neurochem Int 45:987–993
116. Smith CP, Radziszewski P, Borkowski A, Somogyi GT, Boone TB, Chancellor MB (2004) Botulinum toxin a has antinociceptive effects in treating interstitial cystitis. Urology 64: 871–875
117. Giannantoni A, Porena M, Costantini E, Zucchi A, Mearini L, Mearini E (2008) Botulinum A toxin intravesical injection in patients with painful bladder syndrome: 1-year followup. J Urol 179:1031–1034
118. Kuo HC, Chancellor MB (2009) Comparison of intravesical botulinum toxin type A injections plus hydrodistention with hydrodistention alone for the treatment of refractory interstitial cystitis/painful bladder syndrome. BJU Int 104:657–661
119. Chuang YC, Kim DK, Chiang PH, Chancellor MB (2008) Bladder botulinum toxin A injection can benefit patients with radiation and chemical cystitis. BJU Int 102:704–706
120. Gassner HG, Sherris DA (2003) Chemoimmobilization: improving predictability in the treatment of facial scars. Plast Reconstr Surg 112:1464
121. Khera M, Boone TB, Smith CP (2004) Botulinum toxin type A: a novel approach to the treatment of recurrent urethral strictures. J Urol 172:574–575
122. Kuo HC (2003) Effect of botulinum A toxin in the treatment of voiding dysfunction due to detrusor underactivity. Urology 61:550–554
123. Schulte-Baukloh H, Knispel HH (2002) Botulinum-A toxin in the treatment of neurogenic bladder in children. Pediatrics 110:420–421
124. Schulte-Baukloh H, Michael T, Schobert J, Stolze T, Knispel HH (2002) Efficacy of botulinum-A toxin in children with detrusor Hyperreflexia due to myelomeningocele: preliminary results. Urology 59:325–327

125. Lusuardi L, Nader A, Koen M, Schrey A, Schindler M, Riccabona M (2004) Minimally invasive, safe treatment of the neurogenic bladder with botulinum-A-toxin in children with myelomeningocele. Aktuelle Urol 35:49–53
126. Romero RM, Rivas S, Parente A, Fanjul M, Angulo JM (2011) Injection of botulinum toxin (BTX-A) in children with bladder dysfunction due to detrusor overactivity. Actas Urol Esp 35:89–92
127. Hoebeke P, De Caestecker K, Vande Walle J, Dehoorne J, Raes A, Verleyen P et al (2006) The effect of botulinum-A toxin in incontinent children with therapy resistant overactive bladder. J Urol 176:328–330
128. Schulte-Baukloh H, Knispel HH, Stolze Y, Weiss C, Michael T, Miller K (2005) Repeated botulinum-A toxin injections in treatment of children with neurogenic detrusor overactivity. Urology 66:865–870
129. Altaweel A, Jednack R, Bilodeau C, Corcos J (2006) Repeated intradetrusor botulinum toxin type A in children with neurogenic bladder due to myelomeningocele. J Urol 175:1102–1105
130. Schulte-Baukloh H, Herholz J, Bigalke H, Miller K, Knispel HH (2011) Results of a BoNT/A antibody study in children and adolescents after onabotulinumtoxin A (Botox®) detrusor injection. Urol Int 87:434–438
131. Do Ngoc Thanh C, Audry G, Forin V (2009) Botulinum toxin A for neurogenic detrusor overactivity due to spinal cord lesions in children: a retrospective study of seven cases. J Pediatr Urol 5:430–436
132. Kajbafzadeh AM, Ahmadi H, Montaser-Kouhsari L, Sharifi-Rad L, Nejat F, Bazargan-Hejazi S (2011) Intravesical electromotive botulinum toxin type A administration—part II: clinical application. Urology 77:439–445
133. Apostolidis A, Dasgupta P, Denys P, Elneil S, Fowler CJ, Giannantoni A, Karsenty G et al (2009) Recommendations on the use of botulinum toxin in the treatment of lower urinary tract disorders and pelvic floor dysfunctions: a European consensus report. Eur Urol 55:100–119

Chapter 6
Clinical Use of Botulinum Neurotoxins: Pain

Bahman Jabbari and Duarte G. Machado

Abstract Animal data have shown that botulinum neurotoxins (BoNTs) inhibit the release of pain neurotransmitters/neuromodulators (glutamate, substance P, calcitonin-gene-related peptide) and pro-inflammatory agents (prostaglandins, bradykinin, histamine) from peripheral nerve endings and sensory ganglia and reduce the phenomena of peripheral and central sensitization, major factors for pain chronicity. A review of class I and II studies (double blind, placebo controlled) using the criteria set forward by the Therapeutics and Technology Assessment Subcommittee of the American Academy of Neurology shows different levels of efficacy for a large number of human pain disorders: There exists level A evidence (two or more class I studies—established efficacy) for pain of cervical dystonia, chronic migraine and chronic lateral epicondylitis and level-B evidence (one class I or two class II studies—probably effective) for postherpetic and posttraumatic neuralgia, pain of plantar fasciitis, piriformis syndrome and pain in total knee arthroplasty. Level C evidence (one class II study—possibly effective) denotes allodynia of diabetic neuropathy, chronic low back pain, painful knee osteoarthritis, anterior knee pain with vastus lateralis imbalance, pelvic pain, postoperative pain in children with cerebral palsy after adductor hip release surgery, postoperative pain after mastectomy and sphincter spasms and pain after hemorrhoidectomy. The myofascial pain syndrome and chronic daily headaches have level U evidence (efficacy not proven due to controversial results). Results of BoNT treatment trials in episodic migraine and chronic tension headaches justify level A evidence for treatment failure. The end of each assessed category includes a medical comment and suggestions for improvement of future studies. For certain pain syndromes, figures are provided to illustrate the suggested number and site of injections and the appropriate doses.

Keywords Botulinum neurotoxin · Migraine · Neuropathic · Headache · Analgesic · Neuralgia · Plantar fasciitis · Myofascial pain

B. Jabbari (✉) · D. G. Machado
Department of Neurology, Yale University School of Medicine,
LCI 916, 15 York Street, New Haven, CT 06520, USA
e-mail: bahman.jabbari@yale.edu

K. A. Foster (ed.), *Clinical Applications of Botulinum Neurotoxin,*
Current Topics in Neurotoxicity 5, DOI 10.1007/978-1-4939-0261-3_6,
© Springer Science+Business Media New York 2014

6.1 Introduction

Chronic pain is a common medical complaint, and the management of refractory pain is a huge financial burden to the economy. Despite current availability of a large number of analgesic drugs, management of chronic pain is still a challenge for clinicians. Potent analgesics are often helpful, but side effects and drug interactions limit their clinical utility. Therefore, introduction of new drugs with low side-effect profiles, such as botulinum neurotoxins (BoNTs), is welcomed in the arena of chronic pain management.

BoNTs are used widely in clinical medicine for treatment of spasticity, hyperactive movement and autonomic disorders ([1], Chaps. 3–5 of this volume). In these settings, improvements are believed to result from inhibition of acetylcholine release from presynaptic vesicles via the inhibitory effect of BoNTs upon synaptic proteins [2]. In addition to acetylcholine, it is now increasingly recognized that both types A and B toxins (the two in clinical use) inhibit the release of a wide range of neurotransmitters, many of which are essential for initiation and chronicity of pain. Earlier observation of pain relief following treatment of cervical dystonia (CD) with BoNT type A before improvement of neck posture alerted clinicians to an independent analgesic effect for BoNTs. This observation, along with emerging animal data, led to an explosion of clinical trials with BoNTs for pain management in the past two decades. More recently, the discovery of recombinant toxins (chimeras) as a novel analgesic provided a formulation with a potential to retarget, specifically, the sensory neurons for pain treatment [3].

In this chapter, we will first discuss the data derived from animal studies and the mechanisms suggested for the analgesic effect of BoNT administration. Using the efficacy evaluation criteria of the American Academy of Neurology (AAN) [4], we will then review the evidence for efficacy of BoNTs in human pain disorders. To help clinicians regarding the practical aspects of BoNT therapy for pain management, we provide a brief clinical comment after each section. For some indications (common pain disorders), figures are provided to illustrate location and number of injection sites and the suggested doses for treatment.

6.2 Animal Studies

The anatomy of pain includes a complex system of substrates, the activation of which can lead to pain perception. These substrates consist of peripheral pain receptors, pain-conducting c-fibers, sensory cells in peripheral sensory ganglia and sensory spinal, brainstem, thalamic and cortical neurons. Neurotransmitters and neuromodulators at these levels are crucial to the conduction and perception of pain. In addition, pain chronicity and sustenance depends on mechanisms of peripheral and central sensitization. In the former, persistent exposure to a noxious stimulus leads to tissue accumulation of substance P, calcitonin gene-related peptide (CGRP) and glutamate, the pain modulators which coexist in the nerve terminals [5]. The vasodilation and plasma extravasation caused by these agents lead to release of a number of inflammatory mediators such as histamine, bradykinin, prostaglandin and serotonin, which collectively lead to peripheral sensitization of nerve terminals. Peripheral sensitization enhances the release of glutamate and substance P from spinal cord neurons with resultant central sensitization and heightened perception of pain [6]. At the spinal level, enhanced sensitivity of wide dynamic range (WDR) neurons (caused by peripheral sensitization) is also considered a factor since these sensitized neurons begin to perceive non-nociceptive input as nociceptive [7]. Finally, hyperactivity of the sympathetic nervous system in chronic pain disorders enhances pain and contributes to chronicity (sympathetically maintained pain).

Experimental animal studies have demonstrated that BoNTs work on many levels of the pain system anatomy and that their actions upon pain transmitters and modulators reduce peripheral and central sensitization.

a. *At the peripheral pain receptor level*: Administration of BoNT type A into rat bladder, along with inhibition of acetylcholine release, inhibits adenosine triphosphate (ATP) and purinergic receptors (mediator of sensory excitation) leading to reduction of painful bladder spasms and actual reduction of pain receptors [8].
b. *At the level of sensory cells in peripheral sensory ganglion*: In the spinal sensory ganglion and the trigeminal ganglion, data demonstrate significant inhibitory action upon release of pain transmitters and modulators. This is particularly shown for glutamate, which is believed to be actively involved in development of neuropathic pain [9] and which accumulates in the tissue after peripheral nerve injury. In an elegant experiment, injection of BoNT type A before formalin into the rat paw resulted in significant reduction of tissue glutamate accumulation, which paralleled marked relief of the inflammation-related pain caused by formalin [10]. This response occurred in a dose-dependent fashion. Martinelli et al. [11] reported a similar effect on pain relief and glutamate accumulation with local BoNT type A injection after ligation of the sciatic nerve. The authors further demonstrated promotion of nerve regeneration in the BoNT type A treated group manifested by a local increase in regeneration-associated proteins [division cycle 2 (cd c2) and growth associated protein 43 (GAP-43)] in the sciatic nerve and glial fibrillary acidic protein (GFAP) in Schwann cells. Animals treated with BoNT type A also demonstrated quicker recovery of walking pattern and weight bearing compared to controls.
   Several lines of evidence demonstrate that BoNTs inhibit the release of pain peptides, substance P, bradykinin, CGRP and glutamate in vitro and in vivo from the dorsal root and trigeminal ganglia and from rat bladder tissue after injury [12]–[14]. Also, BoNT inhibits a family of G proteins including Rho guanosine triphosphatase which is essential for activation of interleukin-1, an important pro-inflammatory cytokine [15]. Intraprostatic injection of BoNT type A inhibits cyclooxygenase-2 expression and suppresses capsaicin-induced prostatitis in the animal model [16]. Collectively, these observations indicate that BoNTs are capable of reducing peripheral sensitization in chronic pain conditions by alleviating neurogenic inflammation. Finally, BoNT type A impairs sympathetic transmission and thus can interfere with maintenance of pain via decreasing sympathetic overactivity (sympathetically maintained pain) [17].

c. *At the spinal cord level*: Inhibition of pain-related neuropeptides and cytokines and peripheral sensitization indirectly reduces central sensitization of spinal cord neurons. Furthermore, injection of BoNT type A into rat jaw muscles decreases the electrical discharge of muscle spindles, a major sensory input which can enhance central sensitization in chronic pain via burdening sensitized WRD neurons [18]. Also, in the aforementioned formalin model of pain in rat paw, it was shown that pretreatment with BoNT type A reduces development of fos-positive neurons in lamina I, II, IV, and V of the spinal cord, regions that receive nociceptive input, following formalin administration [10]. Indirect involvement of spinal cord neurons following peripheral injection was suggested in one study which demonstrated that injection of I^{125}−labeled BoNT into one gastrocnemius muscle resulted in increased radioactivity in the ipsilateral sciatic nerve and hemicord of the cat [19]. In rat paclitaxel-induced neuropathy, unilateral subplantar injection of BoNT type A resulted in bilateral improvement of mechanical hyperalgesia [20]. More recently, Back-Rojecky et al. [21] showed more evidence for the central effect of the toxin after peripheral administration. In diabetic rats with bilateral allodynia, unilateral subcutaneous injection of BoNT type A in the allodynic region of one affected limb improved allodynia in both limbs. Furthermore, intrathecal injection of the toxin with a smaller dose produced the same effect. Lastly, femtomolar concentrations of BoNT type A inhibit membrane Na channels in rat central and peripheral neurons [22]. Overactivity of sodium channels plays a pivotal role in at least one model of chronic neurogenic pain, erythromyalgia [23]. Verderio et al. [24] measured the traffic of botulinum toxin A and E in brain synaptosomes. Inhibitory synapses were found resistant to both toxins, and the toxins preferentially inhibited the excitatory neurotransmitters.

6.3 Clinical Evidence in Human Subjects

The clinical evidence in this chapter is defined according to the guidelines of the Therapeutics and Technology Assessment Subcommittee of the AAN [25]. In these guidelines, level A comprises two or more class I studies, B indicates at least one class I or two class II studies and C comprises one class II or two consistent class III studies. Level U refers to unproven evidence, inconsistent results (Table 6.1).

6.3.1 Design of the Review

Class I and class II articles were searched online through PubMed (1966 to the end of March 2011) and OvidSP including ahead-of-print manuscripts.

Currently, five forms of BoNTs are widely marketed and are used for treatment of human subjects. Botox (onabotulinumtoxinA), Xeomin (incobotulinumtoxinA), Dysport (abobotulinumtoxinA), and Prosigne (Chinese toxin) are type A toxins. Myobloc (rimabotulinumtoxinB) is type B. Prosigne is not approved by the Food and Drug Administration (FDA) for use in the USA.

Table 6.1 American Academy of Neurology classification of evidence for therapeutic trials [4]

Class I:	A randomized, controlled clinical trial of the intervention of interest with masked or objective outcome assessment, in a representative population. Relevant baseline characteristics are presented and substantially equivalent among treatment groups or there is appropriate statistical adjustment for differences

The following are also required:

    a. Concealed allocation
    b. Primary outcome(s) clearly defined
    c. Exclusion/inclusion criteria clearly defined
    d. Adequate accounting for dropouts (with at least 80 % of enrolled subjects completing the study) and crossovers with numbers sufficiently low to have minimal potential for bias
    e. For noninferiority or equivalence trials claiming to prove efficacy for one or both drugs, the following are also required[a]:

        1. The standard treatment used in the study is substantially similar to that used in previous studies establishing efficacy of the standard treatment (e.g., for a drug, the mode of administration, dose, and dosage adjustments are similar to those previously shown to be effective)
        2. The inclusion and exclusion criteria for patient selection and the outcomes of patients on the standard treatment are substantially equivalent to those of previous studies establishing efficacy of the standard treatment
        3. The interpretation of the results of the study is based on an observed-cases analysis

Class II:	A randomized, controlled clinical trial of the intervention of interest in a representative population with masked or objective outcome assessment that lacks one criterion a–e class I, above, or a prospective matched cohort study with masked or objective outcome assessment in a representative population that meets b–e class I, above. Relevant baseline characteristics are presented and substantially equivalent among treatment groups or there is appropriate statistical adjustment for differences
Class III:	All other controlled trials (including well-defined natural history controls or patients serving as their own controls) in a representative population, where outcome is independently assessed, or independently derived by objective outcome measurements
Class IV:	Studies not meeting class I, II, or III criteria including consensus or expert opinion

[a]Note that numbers 1–3 in class I are required for class II in equivalence trials. If any one of the three is missing, the class is automatically downgraded to a class III

6.3.2 *Pain Disorders with Level A Evidence (Two or More Class I Studies, Efficacy Established)*

6.3.2.1 Neck Pain Associated with CD (Eight Class I Studies)

CD is a late-onset focal dystonia characterized by twisting and twitching of the neck and shoulder muscles. There is often limitation of head movement leading to different head postures: over-rotation (torticollis), lateral tilt (laterocollis), over-flexion (anterocollis) and extension (retrocollis) or a combination thereof. Neck pain is often the most disabling symptom experienced by a majority of the patients (68–75 %) [26].

Eight class I studies evaluated the issue of pain in CD in relation to BoNT treatment. Four investigated type A [27]–[30] and four investigated type B BoNTs [31]–[34]. One other study compared efficacy and safety of abobotulinumtoxinA with trihexyphenidyl [35]. In these studies, the response to pain was measured by different means including a simple pain scoring scale (severe, moderate, mild, none), the visual analog scale (VAS), and the pain subscale of Toronto Western Spasmotic Torticollis Rating Scale (TWSTRS). The results uniformly show that treatment of CD with type A (Botox, Dysport, Xeomin) or type B (Myobloc) BoNTs results in significant reduction of neck pain ($p < 0.05$). For example, in the study of Truong et al. [30] comparing abobotulinumtoxinA with placebo at 4 weeks, the level of pain reduction measured by VAS was 13.4 mm (on a 100-mm scale) for abobotulinumtoxinA versus 1.9 mm for the placebo ($p < 0.002$). AbobotulinumtoxinA is also superior to trihexyphenidyl in terms of efficacy and better tolerance [35].

Additionally, six prospective, blinded, multicenter studies compared two serotypes of BoNTs with each other in terms of safety and efficacy and response to pain [34], [36], [37], [38], [39], [40]. The comparison studies of onabotulinumtoxinA with rimabotulinumtoxinB [34], [36], [37] and incobotulinumtoxinA [38] showed that both serotypes effectively reduced pain and there was no significant difference between the two except in the study of Lew et al. [34], which demonstrated a significantly higher response rate of pain relief for type B (59 % versus 36 %; $p < 0.05$). The comparison study of abobotulinumtoxinA with onabotulinumtoxinA reported slightly more pain improvement in the abobotulinumtoxinA group, but this difference was not statistically significant. In one report, abobotulinumtoxinA group demonstrated more side effects [39]. A recent double-blind class II study compared pain efficacy of onabotulinumtoxinA with Prosigne (using 300 units of each) in patients with CD. Pain efficacy was the same for both toxins at 4 and 16 weeks [40].

Three prospective long-term studies of abobotulinumtoxinA with six or more injections (performed every 3 months) demonstrated sustained responses following repeated treatments with mild side effects (local pain, subtle weakness, dysphagia) [30], [41], [42] Approximately 20 % of the patients chose not to continue the treatment due to high cost, dislike of injections and loss of efficacy [41], [42]

Clinical Comment BoNTs are an effective and established treatment for pain in CD. The degree of pain relief in CD is comparable among type A toxins and is similar between type A and type B toxins (with the exception of one study which reported type B being more effective [34]).

6.3.2.2 Chronic Migraine (Two Class I Studies)

Chronic migraine (CM) is defined as headache with a frequency of 15 or more headache days per month (at least eight migraine type), for more than 3 months, lasting more than four hours per day [43]. Freitag et al. [44], in a double-blind, placebo-controlled study, compared the effect of a fixed dose (100 units), fixed site (glabella, frontalis, temporal, trapezius, suboccipital) paradigm treatment of onabotulinumtoxinA (20 patients) with placebo (21 patients). All patients with medication

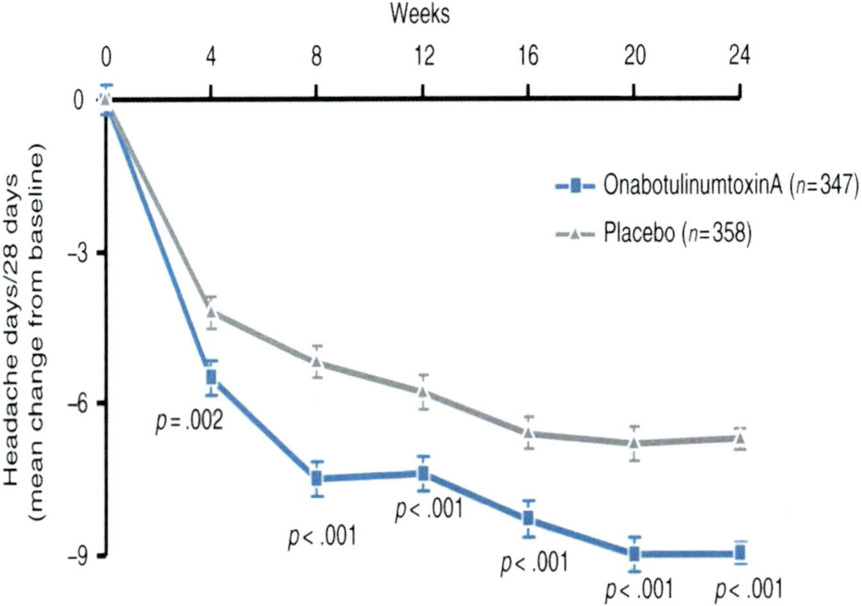

Fig. 6.1 PREEMPT 2: showing significant improvement of pain days from botulinum toxin group compared to placebo group over all time points of the 24-month blinded arm of the study. (From Cephalalgia July 2010 with permission)

overuse were excluded. The primary outcome was the number of migraine episodes experienced over each 4 weeks of the study. The secondary outcomes were number of headache days and headache index (HI; measure of both intensity and frequency). OnabotulinumtoxinA was statistically superior to placebo for both primary ($p < 0.01$) and secondary outcomes (frequency of pain days $p = 0.041$ at 4 weeks and $p = 0.046$ at 16 weeks, and HI, $p = 0.003$ at 16 weeks).

In the summer of 2010, the results of Phase 3 Research Evaluating Migraine Prophylaxis Therapy (PREEMPT) 1 and PREEMPT 2 [45], [46], two large class I, multicenter studies assessing efficacy of onabotulinumtoxinA in CM, were published. Each study included approximately 700 patients, with comparable and close numbers of subjects in both the toxin and placebo groups, evaluated over a 24-week blinded arm study followed by a 32-week open arm study. Both studies included patients with medication overuse. The primary outcome for PREEMPT 1 was the number of headaches episodes, and for PREEMPT 2, the number of headache days, both evaluated at 24 weeks. A number of secondary outcomes were also evaluated at the 24-week time point. PREEMT 2 met its primary and secondary outcomes at all time points (Fig. 6.1 and Table 6.2). For the primary outcome, the change in headache days was 9 for onabotulinumtoxinA versus 6.7 for the placebo ($p < 0.001$). The pooled data [47] of the two studies also showed significant change from the baseline in favor of onabotulinumtoxinA regarding the primary and secondary parameters (Table 6.2).

Table 6.2 Results (*p* values) of PREEMPT studies and pooled data comparing botulinum toxin and placebo with baseline

Parameters	PREEMPT 1	PREEMPT 2	Pooled data
Number of HD days	0.006	< 0.001 (primary outcome)	< 0.001
Number of HD episodes	0.34 (primary outcome)	< 0.003	< 0.001
Number of migraine days	0.002	< 0.001	< 0.001
Number of moderate to severe HD days	0.004	< 0.001	< 0.001
Change in total HIT-6 score	0.001	< 0.001	< 0.001
Total accumulative HD hours in HD days	< 0.001	< 0.001	< 0.001
Frequency of triptane intake	0.23	< 0.001	< 0.001

HD headache, *PREEMPT* phase 3 research evaluating migraine prophylaxis therapy

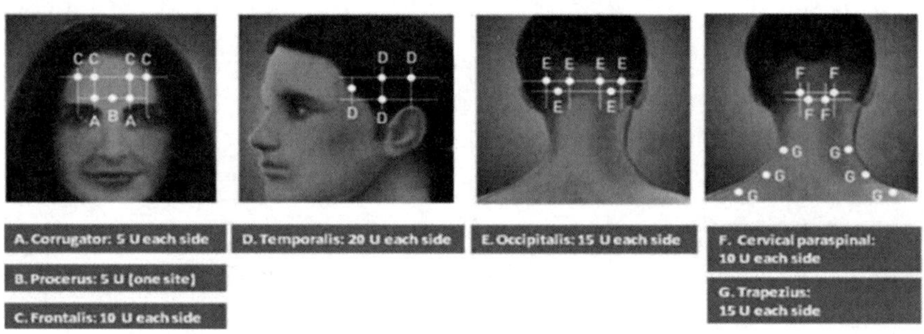

Fig. 6.2 Sites of onabotulinum toxin injection for treatment of chronic migraine based on PRE-EMPT studies (from Blumenfeldt et al Headache 2010;50:146). The recommended dose varies between 155–195 units. With permission Wiley publishers

Although PREEMPT 1 did not meet the primary outcome, it met its secondary outcomes (Table 6.2). The FDA considered headache days a better outcome measure than headache episodes for the study of CM (PREEMPT 2). OnabotulinumtoxinA was approved for treatment of CM in the UK and Canada in the summer of 2010 and in the USA in October 2010. Figure 6.2 shows the site of injections and doses used in the PREEEMPT studies of CM.

Clinical Comment CM is a major health problem and is believed to account for the majority of the cases of chronic daily headaches (CDHs). Many clinicians consider the number of moderate and severe headaches (most troublesome to the patient) a true measure of patient discomfort and a better primary outcome compared to either total number of pain days or headaches episodes. This measure was significant for the toxin group in all three studies (the two PREEMPT studies and the pooled data) (Table 6.2). Clinical evidence in agreement with PREEMPT data (Fig. 6.1) indicates that the analgesic effect of botulinum toxin therapy in CM improves with repeated treatments. Inclusion of patients with medication overuse is considered a weakness of the PREEMPT studies.

6.3.2.3   Chronic Lateral Epicondylitis (Three Class I Studies and One Class II Study)

Wong et al. [48] conducted a prospective, double-blind study in 60 patients with chronic lateral epicondylitis (CLE). In the toxin group, abobotulinumtoxinA (60 units) was injected into subcutaneous tissue and underlying muscle, 1 cm from the lateral epicondyle aimed toward the tender spot. Pain intensity was evaluated by VAS (primary outcome) at 4 and 12 weeks. In the toxin group, pain measured by VAS improved significantly ($p < 0.001$ and $p = 0.006$) for 4- and 12-week time points. One patient developed weakness of fingers, which lasted for 3 months. However, a blinded study of 40 patients with CLE by Hayton et al. [49] found no significant change in VAS or quality of life (measured by the 12-Item Short Form Health Survey or SF-12) 3 months after injection of abobotulinumtoxinA intramuscularly 5 cm distal to the maximum point of tenderness at the lateral epicondyle, in line with the middle of the wrist. In another class I study [50] of 130 patient in 16 centers, BoNT type A was injected in the painful origin of forearm extensor muscle and the results were compared with placebo at 2, 6, 12, and 18 weeks. Both VAS and global assessments improved significantly from week 2 to week 18 at different time points ($p = 0.003$ and 0.001, respectively). Weakness of the third finger developed in the number of patients but it did not interfere with work. In a recent class I study, 48 patients randomly received abobotulinumtoxinA (60 units) or placebo under a double-blind, prospective protocol [51]. The site of injection was one-third of the way down the length of the forearm from the tip of the lateral epicondyle along the course of the posterior interosseous nerve. Primary outcome was improvement of pain at rest (measured by VAS) and secondary outcomes were improvement of pain at maximum grip and maximum pinch. Outcomes were measured at 4, 8 and 16 weeks. Significant improvement of pain at rest and pain at maximum pinch was noted in the BoNT group ($p < 0.01$). Approximately half of the patients in the BoNT group developed pain and muscle spasms in the injected site. One patient developed significant weakness of the third and the fourth finger which lasted for 2 months.

Clinical Comment   The three class I studies with larger number of patients depicted efficacy of BoNT treatment in CLE. The study of Hayton et al., which disclosed negative results, had two possible design problems: (1) The first assessment was done at 3 months, which may be too late since most patients who receive BoNT treatment show fading of improvement by 3 months. (2) The small sample size of the study could have led to type II error in statistical assessment. The side effects, weakness of fingers and wrist extension, limit the practical value of BoNT therapy in CLE. Future studies may consider smaller doses and more refined techniques to avoid this side effect.

6.3.3   Pain Disorders with Level-B Evidence (One Class I or Two Class II Studies): Probably Effective, Should Be Considered for Treatment

6.3.3.1   Postherpetic and Posttraumatic Neuralgia with Allodynia (Each One Class I Study)

Neuropathic pain is a symptom of damage or dysfunction of the peripheral or central nervous systems, and in some cases it may result from nociceptive injury [52]. The pain often has a burning quality and may be associated with dermal hypersensitivity and allodynia. Xiao et al. [53] assessed pain relief by VAS at 1, 7 and 90 days in a class I study in 60 patients with postherpetic neuralgia (PHN) after administering BoNT type A, lidocaine or a placebo (20 patients in each group). Pain relief and improvement of sleep in the BoNT group were superior to that in the lidocaine and placebo groups ($p < 0.05$). Patients in the BoNT group also used significantly less opioids (22 % versus 52 % and 66 %). Ranoux et al. [54] conducted a double-blind, placebo-controlled study on 29 patients with refractory neuropathic pain, 25 with posttraumatic neuralgia (PTN)/allodynia and 4 with PHN. OnabotulinumtoxinA (20–190 units) and placebo were injected once intradermally in the painful area after baseline assessments. Outcomes were evaluated at 4, 12 and 14 weeks with measurement of pain intensity, thermal and mechanical perception, allodynia to skin brushing and quality of life. Patients who received BoNT type A had diminished pain intensity, neuropathic symptoms and allodynic brush sensitivity and reduced number of pain paroxysms along with improvement of certain quality-of-life markers (general activity, mood) compared to the placebo group ($p < 0.05$).

6.3.3.2   Plantar Faciitis (Two Class II Studies)

Pantar faciitis (PF) is the most common cause of heel pain caused by micro-tears and inflammation as a result of repeated injury. In severe cases, treatment with posterior night splints, ultrasound, iontophoresis, phonophoresis, extracorporal shock therapy or local corticosteroid injections can help, but failures are not uncommon. Babcock et al. [55] investigated the efficacy of onabotulinumtoxinA in 27 patients (43 heels) with chronic PF (class II). Injection of 40 and 30 units of onabotulinumtoxinA, one medial to the heel and the other about 1–3 inches anterior to the heel (tender area in PF) (Fig. 6.3), resulted in significant improvement of the pain in the onabotulinum-toxinA group. Two months post injection, the study met all three primary outcomes (reduction of pain intensity measured by pressure algometry, pain frequency and the Maryland Foot Score) ($p < 0.05$).

Huang et al. [56] conducted a prospective, double-blind study in 50 patients with PF and refractory pain. In the toxin group, 50 units of onabotulinumtoxinA were administered into the heel under ultrasonic guidance. At 3 weeks and 3 months, the toxin-injected group showed significant pain relief (measures by VAS) compared to

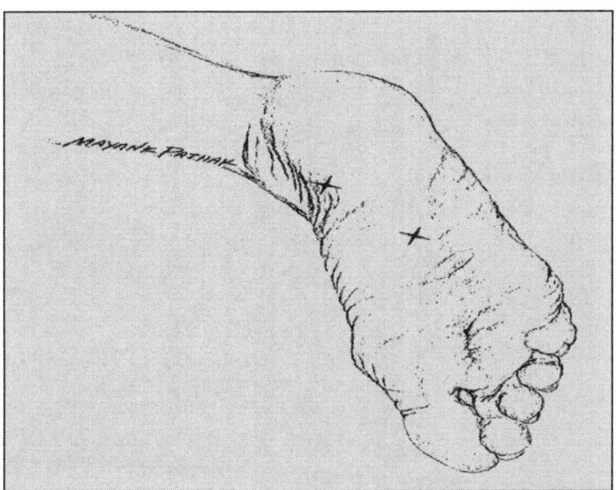

Fig. 6.3 Sites of BoNT-A injections for plantar faciitis (30 and 40 units). [99]

the placebo group ($p < 0.001$). The toxin-treated group also showed improved gait at 3 months as measured by increased center of pressure velocity ($p < 0.05$).

6.3.3.3   Piriformis Syndrome (Two Class II Studies)

The piriformis muscle originates from the anterior part of the sacrum and sacroiliac capsule and after exiting from the pelvis attaches to the greater trocanter. Spasms of the piriformis muscle cause pain deep in the buttock referred to as piriformis syndrome (PS). Childers et al. [57] conducted a double-blind, crossover study in ten patients with PS. OnabotulinumtoxinA, 100 units, was injected into the piriformis muscle under electromyographic and fluoroscopic guidance. The pain relief (measured by VAS scores) was significant in the onabotulinumtoxinA arm of the study compared to the placebo arm ($p < 0.05$). Fishman et al. [58] compared the results of 200 units of onabotulinumtoxinA with lidocaine and steroid injection and with placebo injection into the piriformis muscle in 72 patients with PS; 50 % or better improvement in VAS score was considered significant. Onabotulinumtoxin A was superior to the placebo ($p = 0.001$) and to steroids + lidocaine ($p < 0.005$) in relieving pain.

6.3.3.4   Refractory Painful Total Knee Arthroplasty (One Class I Study)

Refractory pain after total knee arthroplasty (TKA) is common and affects 8–13 % of the patients after surgery [59]. Singh et al. [60] assessed the efficacy of an intra-articular injection of 100 units of onabotulinumtoxinA in 54 patients with TKA. The primary end point was a two grade or more reduction of pain in VAS 2 months after treatment, and secondary end points included Physician's Global Assessment of Change (PGAC), 36-Item Short Form Health Survey (SF-36) and several other scales.

At 2 months, a significant response in VAS was noted in 71 % of the patients in BoNT versus 36 % in the placebo group ($p = 0.025$). Both PGAC and SF-36 (pain subscale) showed significant change in favor of the onabotulinumtoxinA group ($p = 0.003$ and $p = 0.049$, respectively).

Clinical Comment Larger class I studies are necessary to establish the efficacy of BoNT treatment in these painful disorders. Refinement of the technique and dose optimization could potentially lead to better results.

6.3.4   Pain Disorders with Level C Evidence (One Class II Study)

Recommendation: Possibly effective. May be used at the discretion of the clinician.

6.3.4.1   Refractory Low Back Pain

Low back pain (LBP) is the most common form of pain in adults producing some form of disability in 60 % of the patients. Foster et al. [61] studied 31 patients mostly with chronic spine disease (stenosis, disc degeneration) and LBP of more than 6 months duration (class II). They used a fixed paradigm of five lumbar-level injections (L1–L5) with onabotulinumtoxinA, each level receiving 40 units into erector spinae. Primary and secondary outcomes of pain intensity (VAS) and activities of daily living (ADLs) were met and were significantly different from placebo at both 3 weeks and 2 months. At 2 months, 60 % of the patients reported 50 % or more decrease in pain intensity with improvement of at least two ADLs. The same group of investigators conducted a prospective study of 14 months' duration in chronic LBP using the same technique and rating scales (plus a pain frequency scale) [62]. At 2 months, 52 % of the patients showed a significant improvement in all scales compared to placebo. Doses ranged from 250 to 400 units per session. Of early responders, 91 % continued to demonstrate the favorable response with repeat injections. Three patients experienced mild, transient, flu-like reactions.

Clinical Comment LBP has a number of causes which may respond differently to BoNT treatment. The class II study cited above included a heterogeneous group with predominantly unilateral LBP. Selective studies are needed with focus on different causes of LBP and in patients with bilateral LBP.

6.3.4.2   Diabetic Neuropathy

In a double-blind crossover study, Yuan et al. [63] studied the effect of onabotulinumtoxinA versus normal saline subcutaneous administration in 18 patients with diabetic neuropathy. Allodynia and pain sensitivity were assessed by VAS at 1, 4, 8 and 12 weeks. At all time points, onabotulinumtoxin A was superior to saline in reducing pain ($p < 0.05$).

Clinical Comment Study limitation includes the small number of patients. A double-blind study with larger numbers is needed to support the result of this crossover designed study.

6.3.4.3   Painful Knee Osteoarthritis (One Class II Study)

Intra-articular injection of low-dose BoNT type A (100 units), high-dose BoNT type A (200 units), and corticosteroids was investigated in 60 patients, randomly divided into three groups [64]. The primary outcome, significant improvement of VAS at 2 months, was met only for the low-dose BoNT group ($p = 0.01$). All three groups showed a statistically significant response to the secondary outcome, in McMaster Arthritis Index scores for pain, stiffness and function.

Comment One limitation of the study is the large number of dropouts (48 %). It is also hard to explain why the-low dose group fared better than the higher dose group.

6.3.4.4   Anterior Knee Pain Associated with Vastus Lateralis Imbalance

Investigators of this study injected abobotulinumtoxinA (500 units) or saline (1 cc) randomly into the vastus lateralis muscle of 24 patients with anterior knee pain [65]. The primary outcomes, improvement in knee pain-related disability and activity-related knee pain (in VAS) at 3 months, were both met ($p < 0.04$ for disability and < 0.003, < 0.02 and < 0.04 for pain in kneeling, squatting and walking, respectively).

6.3.4.5   Pelvic Pain

Chronic pelvic pain affects 3.8 % of women and imposes an annual burden of approximately US\$ 2 billion (direct and indirect costs) to the US economy. In a double-blind, placebo-controlled study, Abbott et al. [66] investigated the effect of 80 units of onabotulinumtoxin A injected into pelvic floor muscles in 60 women with chronic (> 2 years) pelvic pain and pelvic floor spasms. Pelvic pain was assessed by VAS and pelvic floor pressure was gauged by vaginal manometry monthly for 6 months. Those patients who were injected with onabotulinumtoxinA reported significant relief from nonmenstrual pain compared to the placebo group ($p = 0.009$). The onabotulinumtoxinA group also demonstrated a significant decrease in the pelvic floor pressure ($p < 001$).

Comment Future studies should provide clearer definitions of primary and secondary outcomes.

6.3.4.6   Postoperative Pain in Children with Cerebral Palsy After Adductor Hip Release Surgery

Barwood et al. [67], in a randomized, double-blinded study, reported significant alleviation of postoperative pain in 16 children with cerebral palsy who received BoNT type A injections into thigh adductors before adductor hip release surgery for prevention of hip dislocation ($p < 0.003$). There was also a significant reduction in mean analgesic requirement ($p < 0.05$) and mean length of hospitalization ($p < 0.003$).

6.3.4.7   Postoperative Pain After Mastectomy

In a randomized and placebo-controlled study [68] of 48 patients, injection of 100 units of BoNT type A into the pectoralis major, serratus anterior and rectus abdominis muscles before mastectomy reduced postoperative pain significantly ($p < 0.0001$) and facilitated reconstruction with a tissue expander. The placebo group used more narcotics to alleviate pain postoperatively compared to the BoNT type A group ($p < 0.0001$).

6.3.4.8   Sphincter Spasms and Pain After Hemorrhoidectomy

In a double-blind study [69] of 50 patients, injection of 20 units of BoNT type A into the internal rectal sphincter prior to hemorrhoidectomy resulted in significant reduction of postoperative sphincter spasms ($p < 0.05$).

6.3.5   Pain Disorders with Level-U Evidence: The Evidence to Support or Refute Efficacy Is Insufficient Due to Contradictory Results

6.3.5.1   Myofascial Pain Syndrome

Myofascial pain syndrome (MFPS) is characterized by the presence of focal regions of muscle tenderness and trigger points (tPts) which, upon pressure, provoke radiating pain. The tPts probably represent erratic or dysfunctional motor end plates with excessive acetylcholine content. Table 6.3 summarizes the results of class I and II studies with BoNT treatment in MPS [70]–[78]. As can be seen in this table, each one of the nine studies used different doses per tPt, and responses were evaluated at different time points and with different scales. All studies used BoNT type A toxin, seven onabotulinumtoxinA and one abobotulinumtoxinA. Four studies (including one class I) reported significant pain relief at some point after treatment (two at primary outcome time point), whereas five did not.

Table 6.3 Randomized, controlled trials of botulinum toxins in MFPS

Author	No	Study	Location	Outcome measures	Dose	Result
Freund and Schwartz 2000 [70]	26	Class II	Neck	PO, VAS, ROM, at 4 weeks	B: 20 U/tp	$p = 0.001$
Wheeler et al. 2000 [71]	50	Class II	Cervico-thoracic	PO, NPAD, GAI, SF-36	B: $231 \pm 50\,\mu$	ns
Ferrante et al. 2005 [72]	142	Class II	Neck and shoulder	PO, VAS, PPT, SF-36	B: 10, 25, 50 U/tp	ns
Ojala et al. 2006 [73]	31	Class II	Neck and shoulder	PO, VAS, VRS, PPT at 4 weeks	B: 15–35 U 5 U/tp	ns
Gobel et al. 2006 [74]	144	Class I	Upper back	PO: mild or no pain at 5 weeks	D: 400 U 40 U/tp	$p = 0.002$
Qerma et al. 2006 [75]	30	Class II	Infra-spinatus	PO: pain intensity 0–10 scale (3 and 28 weeks)	B: 50 U/tp 12.5 U/tp	ns
Lew et al. 2007 [76]	29	Class II	Cervico-thoracic	PO: VAS, NDI, SF-36 at 2 months	B: 100–200 U 50 U/tp	ns
Miller et al. 2009 [77]	47	Class II	Cervico-thoracic	PO: VAS, PF at 2 months	B: $150–300\,\mu$	$p = 0.001$ (VAS)
Benecke et al. 2011 [78]	153	Class II	Cervico-thoracic	PO: percent of pts with mild or no pain at 5 weeks	D: 400 U	5 weeks = ns 9 and 10 weeks, $p = 0.04$

B onabotulinumtoxinA, *D* dysport, *GAI* global assessment of improvement, *NDI* neck disability index, *NPAD* neck pain and disability scale, *NS* not significant, *PF* pain frequency, *PO* primary outcome measure, *PPT* pain pressure threshold, *ROM* range of motion, *SF-36* 36-item short form health survey, *tp* trigger point, *VAS* pain intensity in visual analog scale, *VSR* verbal reporting score

Clinical Comment It is not possible at this time to make a firm statement regarding the role of BoNT treatment in MFPS due to the diverse nature of the studies. In positive studies of Gobel et al. [74] and Miller et al. [77], the investigators injected a larger number of tPts (> 5). The negative results of Ferrante et al. [72] might have been be confounded by exclusion of patients with more than five tPts; the cohort probably had a milder form of MFPS. In the study of Bencke et al. [78], pain relief was achieved at 9 and 10 weeks but not at 5 weeks. The fixed pattern of injection might have contributed to earlier pain relief. In the study of Ojala et al. [73], the dose per tPt (5 units) might have been too small to be effective. Future studies of MFPS should use methodologies which have proved effective in the past, perhaps with larger doses and with customized rather than fixed designs.

6.3.5.2 Chronic Daily Headaches

Four class I and II studies addressed the issue of CDH directly. All four class I studies [79]–[82] used a mean change in headache-free days/month as the primary outcome. Three used a flexible injection paradigm [79]–[81]. In one study [79], BoNT type A (200 units) increased the number of headache-free days/month significantly (11 days versus 8 days of placebo) ($p < 0.05$). In another study [80] of 355 patients, the response to BXT type A was compared to placebo over a 9-month period during which the patients received three treatment cycles (105–260 units). The study did not meet the primary outcome. The third study [81] looked at a subset of this cohort, 228 patients with no prophylactic medications. When compared with placebo, the between-group difference was statistically significant at successive time points (for the first 3 months, $p = 0.004$, $p = 0.032$ and $p = 0.023$, respectively). In the fourth study [82], 702 patients were stratified into four groups, one placebo group and three treatment groups (75, 150 and 225 units) with a fixed injection paradigm. The primary outcome measure (an increase in pain-free days) was not met.

Clinical Comment Inconsistent results of the aforementioned studies led to depiction of U evidence for BoNT treatment of CDH by the AAN subcommittee in 2008 [83]. It is fitting to consider each major category of CDH separately, namely CM and chronic tension headaches (CTHs). As mentioned above, the new data illustrated a positive response to BoNT treatment for CM (level-A evidence).

6.3.6 Major Pain Disorders with Predominantly Negative Results: Episodic Migraine and CTHs

6.3.6.1 Episodic Migraine (Four Class I and Four Class II Studies)

The first class I study [84] compared BoNT type A to placebo in 232 patients, each with four to eight episodes of migraine per month. Up to 25 units BoNT type A was injected into the frontal and temporal muscles. Both groups showed a reduction in frequency, intensity, and duration of migraine headaches but the difference between two groups was not statistically significant (at 1 and 3 months). Another class I study [85], investigated the efficacy and safety of BoNT type A in 418 patients with the same migraine frequency using doses of 7.5–50 units. Both BoNT type A and placebo decreased the migraine frequency from baseline at each time point between 1 and 4 months after injection. Again, the difference between the two treatments was not significant. A third class I study [86] enrolled 369 patients, each with 4 to 15 episodes of migraine/month. The patients were stratified into three treatment groups. The total dose of BoNT type A ranged from 110 to 260 units (mean 190 units). The primary outcome was a decrease in frequency of migraine episodes from baseline between days 30 and 180 post treatment. The primary outcome was not met but patients who had the highest pain frequency (12–15 per month) responded considerably better to BoNT type A than to the placebo ($p = 0.041$). The fourth class I study [87] evaluated the efficacy and safety of BoNT type A in 495 patients after a 30-day placebo run-in. Patients were studied in four groups, three on BoNT type A (225 units, 150 units, 75 units) and one on placebo. The primary outcome, frequency of migraine episodes on day 180, was not met for any of the three groups.

The first class II study [88] investigated the effect BoNT type A administration (25 and 75 units) into glabellar and frontal muscles. The primary outcome was the proportion of the patients with 50 % or more reduction of headaches frequency as compared to baseline. This outcome was not met but the BoNT type A group showed a significant decrease in frequency of moderate and severe headaches at 2 months and of any migraine at 3 months ($p < 0.05$). The second class II study [89] compared the effect of two doses of 16 and 100 units of BoNT type A with placebo. The primary outcome, a change in frequency of moderate or severe headaches per month, was not met. The study, however, showed a significant decrease in the proportion of the patients experiencing a reduction of two or more headaches per month. The third class II study [90] also did not find a significant difference in the frequency and severity of episodic migraine (EM) between BoNT type A and placebo after the first of a series of treatments. From the second treatment on, however, the migraine index (frequency $\times$ intensity) was significantly lower for the BoNT type A group at all measured time points. The fourth class II study [91] compared the effect of BoNT type A and divalproex sodium with saline and divalproex sodium in 59 patients with EM and CM. Several primary outcomes, including a decrease in frequency, intensity and disability assessment score, were met for both groups at multiple time points (1, 3, and 6 months). There was, however, no statistically significant difference between the responses of the two groups at any time point.

6.3.6.2   Chronic Tension Headaches

Four class I studies [92]–[95] (two using onabotulinumtoxinA and two using abobotulinumtoxinA) and three class II studies [96]–[98] (one using onabotulinumtoxinA and two using abotulinumtoxinA), investigated the efficacy of BoNT treatment in patients with CTHs. The dose of onabotulinumtoxinA varied from 20 to 150 units and that of abobotulinumtoxinA from 30 to 500 units. Although some secondary outcomes were met, all four class I and two of three class II [96]–[97] studies did not meet their primary outcome which, for most, was the number of pain-free days.

Clinical Comments All class I studies of EM (less than 15 episodes per month) and CTH have shown no improvement with BoNT-A treatment hence denoting a probably ineffective, level A evidence. However, there are important technical issues that need to be discussed and clarified:

1. EM studies have taken frequency of migraine episodes as a primary outcome. This is probably an unrealistic measure since what is most disturbing to the patient is the episodes of moderately severe and severe headaches. Most patients are not much bothered by subtle and mild headache episodes which do not change their quality of life. As discussed above, some studies of EM have shown significance for BoNT treatment in reducing frequency of moderately severe to severe migraine episodes [88] and others have emphasized the importance of migraine severity by showing significant reduction of migraine index (frequency $\times$ severity) in the second treatment [90]. We recommend that future studies of EM take the frequency of moderately severe to severe episodes as the primary outcome measure.

2. The studies of CTH have several limitations:

a. Considering the number of headache-free days (half of the studies) or local skull tenderness (half of the studies) as a primary outcome is probably also unrealistic. The study of Silberstein et al. [94] shows that the BoNT group had a 50 % or more reduction in headaches days ($p = 0.024$) but demonstrated no significant change in headache-free days. A better measure again seems to be number of days with moderate to severe headaches.

b. The majority of CTH studies used a small total dose of the toxin (less than 100 units for onabotulinumtoxinA and less than 500 units for abobotulinumtoxinA), small dose per site, and small number of injected sites. These limitations have been mentioned by the investigators themselves. Recent successful studies of CM (PRE-EMPT II) used a larger number of injection sites, coverage of more muscles and doses larger than 150 units/session (155–195 units). Future studies of CTH with BoNTs may use a technical approach similar to the one which proved effective for CM.

6.4   Conclusion

Over the past decade, BoNT treatment has shown efficacy in a large spectrum of human pain disorders. Animal data have provided evidence for a variety of mechanisms to explain BoNTs' analgesic effects. To date, with the exception of pain in CD, the majority of human pain data comes from investigations conducted with botulinum toxin A and in particular with onabotulinumtoxinA. There is a need for more extensive investigations with other forms of botulinum toxin A and with botulinum toxin B in treatment of pain disorders. Selection of the appropriate primary outcome and proper dosage are crucial for obtaining favorable results in clinical trials with BoNTs in pain disorders.

Acknowledgment Narges Moghimi, M.D, helped with the literature search and with final production of the manuscript.

References

1. Jankovic J, Albanese A, Atassi Z, Dolly O, Hallett M, May N (eds) (2008) Botulinum toxin: therapeutic clinical practice and science. Saunders, New York
2. Blasi J, Chapman ER, Link E, Binz T, Yamasaki S, De Camilli P, Südhof TC, Niemann H, Jahn R (1993) Botulinum neurotoxin A selectively cleaves the synaptic protein SNAP-25. Nature 365:160–163
3. Chaddock JA, Purkiss JR, Alexander FC, Doward S, Fooks SJ, Friis LM, Hall YH, Kirby ER, Leeds N, Moulsdale HJ, Dickenson A, Green GM, Rahman W, Suzuki R, Duggan MJ, Quinn CP, Shone CC, Foster KA (2004) Retargeted clostridial endopeptidases: inhibition of nociceptive neurotransmitter release in vitro, and antinociceptive activity in in vivo models of pain. Mov Disord 19(Suppl 8):S42–47
4. French J, Gronseth G (2008) Lost in a jungle of evidence: we need a compass. Neurology 71:1634–1638
5. Gazerani P, Au S, Dong X, Kumar U, Arendt-Nielsen L, Cairns BE (2010) Botulinum neurotoxin type A (BoNTA) decreases the mechanical sensitivity of nociceptors and inhibits neurogenic vasodilation in a craniofacial muscle targeted for migraine prophylaxis. Pain 151:606–616
6. Aoki KR (2005) Review of a proposed mechanism for the antinociceptive action of botulinum toxin Type A. Neurotoxicology 26:785–793
7. Roberts WJ (1986) A hypothesis on the physiological basis for causalgia and related pains. Pain 24:297–311
8. Lawrence GW, Aoki KR, Dolly JO (2010) Excitatory cholinergic and purinergic signaling in bladder are equally susceptible to botulinum neurotoxin A consistent with co-release of transmitters from efferent fibers. J Pharmacol Exp Ther 334:1080–1086
9. Osikowicz M, Mika J, Makuch W, Przewlocka B (2008) Glutamate receptor ligands attenuate allodynia and hyperalgesia and potentiate morphine effects in a mouse model of neuropathic pain. Pain 139:117–126
10. Cui M, Khanijou S, Rubino J, Aoki KR (2004) Subcutaneous administration of botulinum toxin A reduces formalin-induced pain. Pain 107:125–133
11. Marinelli S, Luvisetto S, Cobianchi S, Makuch W, Obara I, Mezzaroma E, Caruso M, Straface E, Przewlocka B, Pavone F (2010) Botulinum neurotoxin type A counteracts neuropathic pain and facilitates functional recovery after peripheral nerve injury in animal models. Neuroscience 24(171):316–328
12. Dolly JO, Aoki KR (2006) The structure and mode of action of different botulinum toxins. Eur J Neurology 13:1–9
13. Meng J, Wang J, Lawrence G, Dolly JO (2007) Synaptobrevin I mediates exocytosis of CGRP from sensory neurons and inhibition by botulinum toxins reflects their anti-nociceptive potential. J Cell Sci 120:2864–2874
14. Lucioni A, Bales GT, Lotan TL, McGehee DS, Cook SP, Rapp DE (2008) Botulinum toxin type A inhibits sensory neuropeptide release in rat bladder models of acute injury and chronic inflammation. BJU Int 101:366–370
15. Namazi H (2008) Intravesical botulinum toxin A injections plus hydrodistension can reduce nerve growth factor production and control bladder pain in interstitial cystitis: a molecular mechanism. Urology 72:463–464

16. Chuang YC, Yoshimura N, Huang CC, Wu M, Chiang PH, Chancellor MB (2008) Intraprostatic botulinum toxin A injection inhibits cyclooxygenase-2 expression and suppresses prostatic pain on capsaicin induced prostatitis model in rat. J Urol 180:742–748
17. Rand MJ, Whaler BC (1965) Impairment of sympathetic transmission by botulinum toxin. Nature 206:588–591
18. Filippi GM, Errico P, Santarelli R, Bagolini B, Manni E (1993) Botulinum A toxin effects on rat jaw muscle spindles. Acta Otolarynol 113:400–404
19. Wiegand H, Erdmann G, Welhoner HH (1976) 125I-labelled botulinum A neurotoxin: pharmacokinetics in cats after intramuscular injection. Naunyn Schmiedebergs Arch Pharmacol 292:161–165
20. Favre-Guilmard C, Auguet M, Chabrier PE (2009) Different antinociceptive effects of botulinum toxin type A in inflammatory and peripheral polyneuropathic rat models. Eur J Pharmacol 617:48–53
21. Bach-Rojecky L, Salković-Petrisić M, Lacković Z (2010) Botulinum toxin type A reduces pain supersensitivity in experimental diabetic neuropathy: bilateral effect after unilateral injection. Eur J Pharmacol 633:10–14
22. Min-chul S, Watika M, Montomura T, Torri Y, Akaike N (2010) Femtomoar concentratiosn of A type botuinum toxin inhibit membrane NA channels in rat central and peripheral neurons. Poster # 40 presented to IBRCC meeting in Atlanta- October
23. Fischer TZ, Waxman SG (2010) Familial pain syndromes from mutations of the NaV1.7 sodium channel. Ann N Y Acad Sci 1184:196–207
24. Verderio C, Grumelli C, Raiteri L, Coco S, Paluzzi S, Caccin P, Rossetto O, Bonanno G, Montecucco C, Matteoli M, Verderio C, Grumelli C, Raiteri L, Coco S, Paluzzi S, Caccin P, Rossetto O, Bonanno G, Montecucco C, Matteoli M (2007) Traffic of botulinum toxins A and E in excitatory and inhibitory neurons. Traffic 8:142–153
25. Gronseth G, French J (2008) Practice parameters and technology assessments: what they are, what they are not, and why you should care. Neurology 71:1639–1643
26. Comella CL (2008) The treatment of cervical dystonia with botulinum toxins. J Neural Transmission 115:379–583
27. Greene P, Kang U, Fahn S, Brin M, Moskowitz C, Flaster E (1990) Double-blind, placebo controlled trial of botulinum toxin injections for the treatment of spasmodic torticollis. Neurology 40:1213–1218
28. Poewe W, Deuschl G, Nebe A, Feifel E, Wissel J, Benecke R, Kessler KR, eballos-Baumann AO, Ohly A, Oertel W, Kunig G (1998) What is the optimum dose of botulinum toxin A in the treatment of cervical dystonia? Result of double-blind, placebo controlled, dose ranging study using Dysport. German dystonia study group. J Neurol Neurosurg Psychiatry 64:13–17
29. Truong D, Duane DD, Jankovic J, Singer C, Seeberger LC, Comella CL, Lew MF, Rodontzky RL, Danisi FO, Sutton JP, Charles PD, Hauser RA, Sheean GL (2005) Efficacy and safety of botulinum A toxin (Dysport) in cervical dystonia: results of the first US randomized, double-blined, placebo-controlled study. Mov Disord 20:783–791
30. Truong D, Brodsky M, Lew M, Brashear A, Jankovic J, Molho E, Orlova O, Timerbaeva S (2010) Global dysport cervical dystonia study group. Long-term efficacy and safety of botulinum toxin type A (Dysport) in cervical dystonia. Parkinsonism Relat Disord 16:316–323
31. Lew MF, Adornato BT, Duane DD, Dykstra DD, Factor SA, Massey JM, Brin MF, Jankovic J, Rodintzky RL, Singer C, Swenson MR, Tarsy D, Murry JJ, Koller M, Wallace JD (1997) Btulinum toxin type B: a double-blind, placebo-controlled safety and efficacy study in cervical dystonia. Neurology 49:701–707
32. Brin M, F, Lew MF, Adler CH, Comella CL, Factor SA, Jankovic J, O'Brien C, Murry JJ, Wallace JD, Willmer-Hulme A, Koller M (1999) Safety and efficacy of NeuroBloc (botulinum toxin type B) in type-A resistant cervical dystonia. Neurology 53:1431–1438
33. Brashear A, Lew MF, Dystra DD, Comella CL, Factor SA, Rodnitzki RL, Trosch I, Singer C, Brin MF, Murry JJ, Wallace JD, Willmer-Hulme A, Koller M (1999) Safety and efficacy of Neurobloc (Botulinum toxin type B) in type-A responsive cervical dystonia. Neurology 53:1439–1446

34. Lew MF, Chinnapongse R, Zhang Y, Corliss M (2010) RimabotulinumtoxinB pain associated with cervical dystonia: results of placebo and comparator-controlled studies. Int J Neurosci 120:298–300

35. Brans JW, Linderboom R, Snoek JW, Zwartz MJ, van weerden TW, Brunt ER, van Hilten JJ, van der Kamp W, Prins MH, Speelman JD (1996) Botulinum toxin versus trihexyphenidyl in cervical dystonia: a prospective, randomized, double-blind trial. Neurology 46:1066–1072

36. Comella CL, Jankovic J, Shannon KM, Tsui J, Swenson M, Leurgans S, Fan W (2005) Dystonia study group. Comparison of botulinum toxin serotypes A and B for the treatment of cervical dystonia. Neurology 65:1423–1429

37. Pappert EJ, Germanson T (2008) Myobloc/Neurobloc European cervical dystonia study group. Botulinum toxin type B vs. type A in toxin-naïve patients with cervical dystonia: randomized, double-blind, non-inferiority trial. Mov Disord 23:510–517

38. Benecke R, Jost WH, Kanovsky P, Ruzicka E, Comes G, Grafe S (2005) A new botulinum toxin type A free of complexing proteins for treatment of cervical dystonia. Neurology 64:1949–1951

39. Odergren T, Hajaltason H, Kaakola S, Solders G, Hanko J, Fehling C, Marttila RJ, Lundh H, Gedin S, Westergern I, Richardson A, Dott C, Cohen H (1998) A double blind, randomized, parallel group study to investigate the dose equivalance of Dysport and Onabotulinum toxin A in the treatment of cervical dystonia. J Neurol Neurosurg Psychiatry 64:6–12

40. Quagliato EM, Carelli EF, Viana MA (2010) Prospective, randomized, double-blind study, comparing botulinum toxins type A botox and prosigne for blepharospasm and hemifacial spasm treatment. Clin Neuropharmacol 33(1):27–31

41. Poewe W, Schelosky L, Kleedorfer B, Heinen F, Wagner M, Deusch G (1992) Treatment of spasmodic torticollis with local injections of botulinum toxin. One year follow up in 37 patients. J Neurol 239:21–25

42. Kessler KR, Skutta M, Benecke R (1999) Long-term treatment of cervical dystonia with botulinum toxin A: efficacy, safety and antibody frequency. German dystonia study group. J Neurol 246:265–274

43. Scher AI, Stewart WF, Liberman J, Lipton RB (1998) Prevalence of frequent headache in a population sample. Headache 38:497–506

44. Freitag FG, Diamond S, Diamond M, Urban G (2008) Botulinum toxin type A in the treatment of chronic migraine without medication overuse. Headache 48:201–209

45. Aurora SK, Dodick DW, Turkel CC, DeGryse RE, Silberstein SD, Lipton RB, Diener HC, Brin MF (2010) PREEMPT 1 Chronic Migraine Study Group. OnabotulinumtoxinA for treatment of chronic migraine: results from the double-blind, randomized, placebo-controlled phase of the PREEMPT 1 trial. Cephalalgia 30:793–803

46. Diener HC, Dodick DW, Aurora SK, Turkel CC, DeGryse RE, Lipton RB, Silberstein SD, Brin MF (2010) PREEMPT 2 Chronic Migraine Study Group. OnabotulinumtoxinA for treatment of chronic migraine: results from the double-blind, randomized, placebo-controlled phase of the PREEMPT 2 trial. Cephalalgia 30:804–814

47. Dodick DW, Turkel CC, DeGryse RE, Aurora SK, Silberstein SD, Lipton RB, Diener HC, Brin MF (2010) PREEMPT Chronic Migraine Study Group. OnabotulinumtoxinA for treatment of chronic migraine: pooled results from the double-blind, randomized, placebo-controlled phases of the PREEMPT clinical program. Headache 50:921–936

48. Wong SM, Hui AC, Tong PY, Poon DW, Yu E, Wong LK (2005) Treatment of lateral epicondylitis with botulinum toxin: a randomized, double-blind, placebo-controlled trial. Ann Intern Med 143(11):793–797

49. Hayton MJ, Santini AJ, Hughes PJ, Frostick SP, Trail IA, Stanley JK (2005) Botulinum toxin injection in the treatment of tennis elbow. A double-blind, randomized, controlled, pilot study. J Bone Joint Surg Am 87:503–507

50. Placzek R, Drescher W, Deuretzbacher G, Hempfing A, Meiss AL (2007) Treatment of chronic radial epicondylitis with botulinum toxin A. A double-blind, placebo-controlled, randomized multicenter study. J Bone Joint Surg Am 89:255–260

51. Espandar R, Heidari P, Rasouli MR, Saadat S, Farzan M, Rostami M, Yazdanian S, Mortazavi SM (2010) Use of anatomic measurement to guide injection of botulinum toxin

for the management of chronic lateral epicondylitis: a randomized controlled trial. CMAJ 182:768–773

52. Argoff CE (2002) A focused review on the use of btulinum toxins for neuropathic pain. Clinical J Pain 18:S177–S181

53. Xiao L, Mackey S, Hui H, Xong D, Zhang Q, Zhang D (2010) Subcutaneous injection of botulinum toxin A is beneficial in postherpetic neuralgia. Pain Med 11:1827–1833

54. Ranoux D, Attal N, Morain F, Bouhassira D (2008) Botulinum toxin type A induces direct analgesic effects in chronic neuropathic pain. Ann Neurol 64:274–283

55. Babcock MS, Foster L, Pasquina P, Jabbari B (2005) Treatment of pain attributed to plantar fasciitis with botulinum toxin A: a short-term, randomized, placebo-controlled, double-blind study. Am J Phys Med Rehabil 84:649–654

56. Huang YC, Wei SH, Wang HK, Lieu FK (2010) Ultrasonographic guided botulinum toxin type A treatment for plantar fasciitis: an outcome-based investigation for treating pain and gait changes. J Rehabil Med 42:136–140

57. Childers MK, Wilson DJ, Gnatz SM, Conway RR, Sherman AK (2002) Botulinum toxin type A use in piriformis muscle syndrome: a pilot study. Am J Phys Med Rehabil 81:751–759

58. Fishman LM, Anderson C, Rosner B (2002) BOTOX and physical therapy in the treatment of piriformis syndrome. Am J Phys Med Rehabil 81:936–942

59. Brander VA, Stulberg SD, Adams AD, Harden RN, Bruehl S, Stanos SP, Houle T (2003) Predicting total knee replacement pain: a prospective, observational study. Clin Orthop Relat Res 416:416–427

60. Singh JA, Mahowald ML, Noorbaloochi SJ (2010) Intraarticular botulinum toxin a for refractory painful total knee arthroplasty: a randomized controlled trial. Rheumatol 37:2377–2386

61. Foster L, Clapp L, Erickson M, Jabbari B (2001) Botulinum toxin A and chronic low back pain: a randomized, double-blind study. Neurology 56:1290–1293

62. Jabbari B, Ney J, Sichani A, Monacci W, Foster L, Difazio M (2006) Treatment of refractory, chronic low back pain with botulinum neurotoxin A: an open-label, pilot study. Pain Med 7:260–264

63. Yuan RY, Sheu JJ, Yu JM, Chen WT, Tseng IJ, Chang HH, Hu CJ (2009) Botulinum toxin for diabetic neuropathic pain: a randomized double-blind crossover trial. Neurology 72:1473–1478

64. Boon AJ, Smith J, Dahm DL, Sorenson EJ, Larson DR, Fitz-Gibbon PD, Dykstra DD, Singh JA (2010) Efficacy of intra-articular botulinum toxin type A in painful knee osteoarthritis: a pilot study. PM & R 2:268–276

65. Singer BJ, Silbert PL, Song S, Dunne JW, Singer KP (2011) Treatment of refractory anterior knee pain using botulinum toxin type A (Dysport) injection to the distal vastus lateralis muscle: a randomised placebo controlled crossover trial. Br J Sports Med 45:640–645

66. Abbott JA, Jarvis SK, Lyons SD, Thomson A, Vancaille TG (2006) Botulinum toxin type A for chronic pain and pelvic floor spasm in women: a randomized controlled trial. Obstet Gynecol 108:915–923

67. Barwood S, Baillieu C, Boyd R, Brereton K, Low J, Nattrass G, Graham HK (2000) Analgesic effects of botulinum toxin A: a randomized, placebo-controlled clinical trial. Dev Med Child Neurol 42:116–121

68. Layeeque R, Hochberg J, Siegel E, Kunkel K, Kepple J, Henry-Tillman RS, Dunlap M, Seibert J, Klimberg VS (2004) Botulinum toxin infiltration for pain control after mastectomy and expander reconstruction. Ann Surg 240:608–613

69. Davies J, Duffy D, Boyt N, Aghahoseini A, Alexander D, Leveson S (2003) Botulinum toxin (botox) reduces pain after hemorrhoidectomy: results of a double-blind, randomized study. Dis Colon Rectum 46:1097–1102

70. Freund BJ, Schwartz M (2000) Treatment of whiplash associated neck pain [corrected] with botulinum toxin-A: a pilot study. J Rheumatol 27:481–484

71. Wheeler AH, Goolkasian P, Gretz SS (1998) A randomized, double-blind, prospective pilot study of botulinum toxin injection for refractory, unilateral, cervicothoracic, paraspinal, myofascial pain syndrome. Spine 23:1662–1666

72. Ferrante FM, Bearn L, Rothrock R, King L (2005) Evidence against trigger point injection technique for the treatment of cervicothoracic myofascial pain with botulinum toxin type. Anesthesiology 103:377–383
73. Ojala T, Arokoski JP, Partanen J (2006) The effects of small doses of botulinum toxin A on neck-shoulder myofascial pain syndrome. A double blind, randomized, and crossover trial. Clin J Pain 22:90–96
74. Göbel H, Heinze A, Reichel G, Hefter H, Benecke R (2006) Dysport myofascial pain study group. Efficacy and safety of a single botulinum type A toxin complex treatment (Dysport) for the relief of upper back myofascial pain syndrome: results from a randomized double-blind placebo-controlled multicentre study. Pain 125:82–88
75. Qerama E, Fuglsang-Frederiksen A, Kasch H, Basch FW, Jensen TS (2006) A double-blind, controlled study of botulinum toxin A in chronic myofascial pain. Neurology 67:241–245
76. Lew HL, Lee EH, Castaneda A, Kilma R, Date E (2008) Therapeutic doses of botulinum toxin A in treating neck and upper back pain of myofascial origin: a pilot study. Arch Phys Med Rehabil 89:75–80
77. Miller D, Richardson D, Eisa M, Bajwa RJ, Jabbari B (2009) Botulinum neurotoxin-A for treatment of refractory neck pain: a randomized, double-blind study. Pain Med 10:1012–1017
78. Benecke, R, Heinze A, Reichel G, Hefter H, Gobel H (2011) Dysport myofascial pain study group. Botulinum type A toxin complex for the relief of upper back myofascial. Pain Medicine 12:1607–1614
79. Ondo WG, Vuong KD, Derman HS (2004) Botulinum toxin A for chronic daily headache: a randomized, placebo-controlled, parallel design study. Cephalalgia 24:60–65
80. Mathew NT, Frishberg BM, Gawel M, Dimitrova R, Gibson J, Turkel C (2005) Botulinum toxin type A (BOTOX) for the prophylactic treatment of chronic daily headache: a randomized double-blind, placebo-controlled trial. Headache 45:293–307
81. Dodick DW, Mauskop A, Elkind AH, DeGryse R, Brin MF, Silberstein SD (2005) BOTOX CDH Study Group. Botulinum toxin type a for the prophylaxis of chronic daily headache: subgroup analysis of patients not receiving other prophylactic medications: a randomized double-blind, placebo-controlled study. Headache 45:315–324
82. Silberstein SD, Stark SR, Lucas SM, Christie SN, Degryse RE, Turkel CC (2005) Botulinum toxin type a for the prophylactic treatment of chronic daily headache: a randomized, double blind, placebo controlled trial. Mayo Clin Proc 80:1126–1137
83. Naumann M, So Y, Argoff CE, Childers MK, Dykstra DD, Gronseth GS, Jabbari B, Kaufmann HC, Schurch B, Silberstein SD, Simpson DM (2008) Therapeutics and technology assessment subcommittee of the American academy of neurology. Botulinum neurotoxin in the treatment of autonomic disorders and pain (an evidence-based review). Neurology 70:1707–1714
84. Saper JR, Mathew NT, Loder EW, DeGryse R, VanDenburgh AM (2007) BoNTA-009 study group. A double-blind, randomized, placebo-controlled comparison of botulinum toxin type A injection sites and doses in the prevention of episodic migraine. Pain Med 8:478–485
85. Elkind AH, O'Carroll P, Blumenfeld A, DeGryse R, Dimitrova R (2006) BoNTA-024-026-036 study group. A series of three sequential, randomized, controlled studies of repeated treatments with botulinum toxin type A for migraine prophylaxis. Pain 7:688–696
86. Aurora SK, Gawel M, Brandes JL, Pokta S, Vandenburgh AM (2007) Botulinum toxin type A prophylactic treatment of episodic migraine: a randomized double blind, placebo-controlled exploratory study. Headache 47:486–499
87. Rejala M, Poole AC, Schoenen J, Pascual J, Lei X, Thompson C (2007) European BoNTA headache study group. A multi-center randomized, placebo-controlled, parallel group study of multiple treatments of botulinum toxin type A (BoNTA) for the prophylaxis of episodic migraine headaches. Cephalalgia 27:492–503
88. Silberstein S, Mathew N, Saper J, Jenkins S (2000) Botulinum toxin type A as a migraine preventive treatment. For the BOTOX Migraine Clinical Research Group. Headache 40:445–450
89. Evers S, Vollmer-Haase J, Schwaag S, Rahmann A, Husstedt IW, Frese A (2004) Botulinum toxin A in the prophylactic treatment of migraine-a randomized, double-blind, placebo-controlled study. Cephalalgia 24:838–843

90. Vo AH, Satori R, Jabbari B, Green J, Killgore WD, Labutta R, Campbell WW (2007) Botulinum toxin type-a in the prevention of migraine: a double-blind controlled trial. Aviat Space Environ Med 78(5 Suppl):B113–B118

91. Blumenfeld AM, Schim JD, Chippendale TJ (2008) Botulinum toxin A and divalproex sodium for prophylactic treatment of episodic or chronic migraine. Headache 48:210–220

92. Padberg M, de Bruijn SF, de Haan RJ, Tavy DL (2004) Treatment of tension-type headache with botulinum toxin: a double-blind, placebo-controlled clinical trial. Cephalalgia 24:675–680

93. Schulte-Mattler WJ, Krack P (2004) Treatment of chronic tension- type headache with botulinum toxin A. A randomized, double blind, placebo-controlled multi-center study. Pain 109:110–114

94. Silberstein SD, Gobel H, Jensen R, Elkin AH, Degryse R, Walcott JM, Turkel C (2006) Botulinum toxin type A in the prophylactic treatment of tension- type headache: a multi-center, double blind, randomized, placebo-control, parallel-group study. Cephalalgia 26:790–800

95. Straube A, Empl M, Ceballos-Baumann A, Tolle T, Stefenelli U, Pfaffenrath V (2008) Dysport tension-type headache study group. Pericranial injection of botulinum toxin type A (Dysport) for tension-type headache- a multi-center, double blind, randomized, placebo-controlled study. Eur J Neurol 15:205–213

96. Rollnik JD, Tanneberger O, Shubert M, Schnider U, Dengler R (2000) Treatment of tension-type headache with botulinum toxin A: a double-blind, placebo-controlled study. Headache 40:300–305

97. Schmitt WJ, Slowey E, Fravi N, Weber S, Burgunder JM (2001) Effect of botulinum toxin A injections in the treatment of tension-type headaches: a double- blind, placebo-controlled trial. Headache 41:658–664

98. Harden RN, Cottrill J, Gagnon CM, Smitherman TA, Weinland SR, Tann B, Joseph P, Lee TS, Houle TT (2009) Botulinum toxin A in the treatment of chronic tension-type headache with cervical myofascial trigger points: a randomized, double-blind, placebo-controlled pilot study. Headache 49:732–743

99. Jabbari B, Babcock MS (2009) Treatment of plantar faciitis with botulinum toxin. In: Truong D, Dressler D, Hallett M (eds) Manual of botulinum toxin therapy. Cambridge University Press, Cambridge, pp 185–188

Chapter 7
Future Developments: Engineering the Neurotoxin

John Chaddock

Abstract Understanding the structure and molecular basis of neurotoxin function has opened up opportunities to engineer novel therapeutic proteins that utilise the neurotoxins and neurotoxin domains. These opportunities and the status of their development are reviewed in this chapter, which brings together the findings detailed in the companion volume to this book, KA Foster (ed) *Molecular Aspects of Botulinum Neurotoxin*, Springer, New York, and shows how they can be applied for the development of innovative therapeutics and research tools.

Keywords Botulinum neurotoxin · Chimera · Domain · Endopeptidase · Targeted secretion inhibitors · Recombinant · Engineering · Delivery

7.1 Introduction

One of the first questions to answer when proposing a topic such as 'Engineering the neurotoxin' is 'Why'? Why is it worth the significant investment in time and resources to alter botulinum neurotoxin (BoNT), the most potent toxin known to man? What innovation or benefits are going to emerge from such studies? Over what timescale will these benefits emerge, and for whom? For those who are not close to the detail of this fascinating class of proteins, these are fair questions to ask and it is up to those of us exploring the toxin and promoting its uses to provide answers and to provide evidence of the benefits. The purpose of this chapter is to consider such questions in light of the preceding chapters and those in the companion volume to this book (KA Foster (ed) *Molecular Aspects of Botulinum Neurotoxin*, Springer, New York) and to provide a commentary that will hopefully inform those inside and outside the field of the huge potential for medical and scientific advancement from this novel protein class.

This chapter (1) briefly reviews the scientific advancements that have been achieved in understanding the relationship between botulinum toxin structure and biological function; (2) considers the opportunities to modify the amino acid building

J. Chaddock (✉)
Syntaxin Ltd., Abingdon, UK
e-mail: john.chaddock@ipsen.com

K. A. Foster (ed.), *Clinical Applications of Botulinum Neurotoxin,*
Current Topics in Neurotoxicity 5, DOI 10.1007/978-1-4939-0261-3_7,
© Springer Science+Business Media New York 2014

blocks of each domain that comprises the neurotoxin and the impact this can have; (3) reviews the progress made towards expanding the medical applications of the toxin through protein engineering; and (4) draws parallels between BoNT and other bacterial toxin fields (such as diphtheria, *Pseudomonas*) and comments on the direction of BoNT research and development. Given the scope of the companion volume to this book (KA Foster (ed) *Molecular Aspects of Botulinum Neurotoxin*, Springer, New York), this chapter will necessarily refer to a number of chapters from that volume for supportive detail and discussion of specific points.

7.2 Structural Advancements That Have Led to Protein Engineering Possibilities

Almost 200 years ago between 1817 and 1822, Kerner published the first accurate and complete descriptions of the symptoms of food-borne botulism and attributed the intoxication to a biological poison. Indeed, Kerner was also the first to suggest that the physiological effects of BoNT could be put to therapeutic use [13]. However, it was not until 1897 when Prof. Emile-Pierre-Marie van Ermengem, a distinguished microbiologist at the University of Ghent, identified an anaerobic bacterium that he termed *Bacillus botulinus* (subsequently classified as *Clostridium botulinum*) that released a potent toxin, now termed BoNT that caused the symptoms associated with food-borne botulism. At this time, the structural organisation of the toxin was essentially unknown. In fact, it took a further 50 years of experimentation, until 1946, before pure crystalline botulinum toxin of serotype A (BoNT/A) was obtained. During the following 20 years, the fractionation of crystalline toxin allowed an understanding of the multi-protein complex nature of the botulinum toxin.

In 1980, Alan Scott published a report on the local injection of BoNT/A into ocular muscles to correct strabismus [27]. This pioneering work initiated a series of investigations into the medical applications of BoNT in a range of neuromuscular conditions. See Chaps. 3 and 6 of this volume for a discussion of this. Twenty years after the first clinical use of BoNT, the publication of the first X-ray structure of BoNT/A was reported [20]. This seminal work established the current view of BoNT structural organisation and 'brought the molecule to life'. As described in detail by Subramanyam Swaminathan in Chap. 5 of the companion volume to this book (KA Foster (ed) *Molecular Aspects of Botulinum Neurotoxin*, Springer, New York), the primary, secondary, tertiary and quaternary structure of the BoNT family of proteins has now been extensively studied. These investigations have facilitated an in-depth awareness of the molecular architecture of this exciting class of proteins and have enabled comparison with other protein toxins and protein domains with similar mechanisms of action. The light chain (LC) metalloendoprotease, for example, shares many features with non-clostridial metalloproteases. Such studies have also started to enable the correlation of structural differences that lead to distinguishing functions. For example, the gross three-dimensional arrangements of BoNT/A and BoNT/B were determined in 1998 [20] and 2000 [30], respectively, and were demonstrated to be essentially similar, a three-domain protein in which the domains are arranged in a linear fashion with the translocation domain in the middle flanked by the binding and catalytic domains. Even though BoNT/A and BoNT/B are comprised of receptor-binding domains that have dissimilar binding targets, and LC domains that cleave different soluble N-ethylmaleimide-sensitive factor attachment protein receptor (SNARE) substrates, the overall tertiary structures of the two neurotoxins were extraordinarily similar.

With little further holotoxin structural data available, and the known sequence similarities between serotypes, it was simple to make the assumption that the tertiary structure of the toxins was likely to be similar across all the serotypes and subtype classes. However, in 2008 the tertiary structure of BoNT/E was determined [18] and, though the individual domains showed a high degree of structural similarity with the A and B cases (indeed, the LC domain of BoNT/E cleaves the same substrate as BoNT/A), there is a major organisational difference between the holotoxin domains. Whereas the three domains are arranged in a linear organisation in A and B, in serotype E the catalytic and binding domains are arranged on the same side of the translocation domain and consequently share interactions that would be absent in A or B. The modified organisation of BoNT/E has been proposed by Kumaran et al. to correlate with the observed more rapid onset of action of BoNT/E, BoNT/E being in a 'translocation-ready' conformation [18]. What these observations inform us is that nature has developed a number of approaches to creating multimeric proteins and we should strive to determine the structural information for all BoNT subtypes.

Described in detail in Chap. 5 of the companion volume to this book, (KA Foster (ed) *Molecular Aspects of Botulinum Neurotoxin*, Springer, New York), structural data from crystallographic studies are now available for (1) the individual catalytic domains (LC) of serotypes A, B, C, D, E, F and G; (2) the binding domain (H_C) of A, B, C, D, F and G; (3) the di-chain LH_N species comprising the LC and the H_N domain of serotypes A and B; and (4) the holotoxins BoNT/A, BoNT/B and BoNT/E. It should be appreciated that, even with this level of advancement, the current level of understanding is not yet fully comprehensive. For example, the structural understanding of the holotoxins is far from complete: In particular, the recently understood alternative conformation of BoNT/E indicates that not all holotoxins should be assumed to have the same overall fold. This could have a significant impact on the biology of the toxins. Similarly, although the LH_N/A and LH_N/B structures indicated a significant similarity to those of the parent neurotoxins, it should not be assumed that this will be the case for all serotypes. As described later in this chapter when discussing the LH_N as the core of the targeted secretion inhibitor (TSI) platform, the precise molecular architecture of the domains is critical to the design of effective engineered proteins.

Understanding the molecular architecture of the individual domains and their interrelationships within the macromolecule has provided an opportunity for at least two paths of protein engineering. Firstly, this knowledge has enabled the rational design of mutants to explore individual domain functions. For example, the relationship between the LC domain and SNARE substrates has been extensively explored and the catalytic mechanism interrogated. Novel LCs have been created that have altered SNARE protein targets [9],[31], leading to the possibility to increase the therapeutic opportunities for the BoNT protein family. Furthermore, structural understanding has been critical in exploring the binding domain and the specific contributions of side chains to the binding event that is so crucial to toxicity. As a consequence of such studies, opportunities to design inhibitors of BoNTs and novel, non-natural proteins with unique properties have both progressed significantly.

7.3 Engineering Improved Neurotoxins

Nature has evolved a range of neurotoxins that are, somewhat artificially, categorized by their reactivity to standard antisera. Within the seven serotypes of toxin are represented a range of subtypes (classified as being $\geq 2.6\%$ difference in amino acid sequence [3]; for example, there are five subtypes of BoNT/A alone. See Chap. 10 of the companion volume to this book (KA Foster (ed) *Molecular Aspects of Botulinum Neurotoxin*, Springer, New York) for further discussion of this. This pool of evolved toxins represents a rich library of proteins that have different properties and characteristics. Whilst nature has provided this range of BoNTs differing in their binding selectivity, SNARE substrate specificity and duration of inhibition of neurotransmitter release, the opportunity exists to use structural understanding to create recombinant variants of the neurotoxins with improved clinical properties. Two of the most obvious domain features for manipulation are the H_C-mediated binding event and the LC-mediated substrate cleavage event. One approach to investigate both engineering opportunities is to create hybrids between different serotype/subtypes of BoNTs and combine the properties of the component domains. For example, Wang and colleagues [32] have reported the creation of chimeras of BoNT/A and /E in which the H_C domain of one serotype was expressed recombinantly fused to the LC and H_N (LH_N) of the other serotype. The translocation properties of the hybrid were clearly differentiated, and they reflected those of the parent LH_N domains, whilst the neuronal specificity was influenced by the identity of the H_C. Such a gross domain-swapping approach is clearly able to harness the inherent biological properties of the domains to create novel entities with unique properties and is the most straightforward approach to BoNT modification. Such an approach to the engineering of the native BoNT structure through domain switching has been further described by Dolly as the basis for the construction of Botulinum neurotoxin enzymatically inactive mutants (BoTIMs) (full-length BoNTs incorporating catalytically inactive LC/A) [11]. By recombinantly combining Botulinum neurotoxin enzymatically inactive mutants (BoTIMs) incorporating LC/E domains, a hybrid protein was constructed that utilised components within the LC/A element to extend the intracellular persistence of the LC/E and therefore the duration of action of LC/E-induced SNAP-25 cleavage [32]. Dolly proposed that the LC/E-induced cleavage of SNAP-25 would be advantageous for specific conditions, for example, in the treatment of various pain states, including chronic pain [4], [11].

Aside from utilising structural and modelling information to make site-specific or domain swap-engineered novel proteins, the field is poised to take further advantage of nature's efforts to create protein variants. Each BoNT serotype actually represents a family of proteins, and as these subtypes are explored for their specific structural and functional properties, the field will become richer in possibilities for the development of novel proteins. As a good example, the BoNT/A family comprises at least five identifiable subtypes and it has recently been demonstrated that the A2 subtype of BoNT/A has a faster onset than the current clinical BoNT/A products based on A1 [24]. Also, the A3 and A4 subtypes of BoNT/A have been observed to possess different catalytic properties to the A1 and A2 subtypes [17]. Further study of subtype biology, in particular, concentrating on the catalytic properties of the LC, the binding properties of the H_C domain, the pH dependency of the H_N domains and the immunogenic contribution of the domains to defining the serotype family will greatly enhance the opportunities for creation of new, useful materials, utilising the broad library that nature has evolved either directly or through the construction of hybrids or through engineering.

The various molecular strategies employed by BoNTs to bind to the cell are starting to emerge and are explored in great detail by Rummel in Chap. 6 of the companion volume to this book (KA Foster (ed) *Molecular Aspects of Botulinum Neurotoxin*, Springer, New York). It is now generally understood that BoNT/B and /G use gangliosides and the intralumenal domain of synaptotagmin as dual receptors, whilst BoNT/A, D, E use gangliosides and the intralumenal domain of synaptic vesicle glycoprotein 2 (SV2). A protein-binding domain has however not yet been identified in BoNT/C suggesting the BoNT/C uses ganglioside only, or an as-yet-unidentified proteinaceous component. Exploring the domain structures in more detail, Rummel [25] identified a ganglioside-binding cavity within the C-terminal sub-domain of the H_C domain (H_{CC}) of BoNT/A and /B defined by the conserved motif H. . . SXWY . . . G. The modification of residues within this site modified both the ganglioside-binding affinity and toxicity of the neurotoxin. By switching the H_{CC} domain of BoNT/B into BoNT/A, Rummel [26] was able to enhance the potency of BoNT/A fourfold. Interestingly, specific modifications to the ganglioside-binding site can both reduce binding and toxicity and enhance it by up to threefold relative to wild-type toxin. This provides a rational basis for engineering mutated BoNT with modified, particularly enhanced, potency as improved clinical products. One such molecule, TrapoX (incorporating a specific mutation of the heavy chain (HC)-binding site), could lead to lower therapeutic dosages of BoNT if it were developed into a product [4], which may provide advantages in minimising off-target effects and maximising tolerance to the protein. As proposed earlier, further investigation into the properties of the BoNT subtypes may well provide additional opportunities in this area.

Manipulation of BoNTs through the creation of hybrids is a powerful technique and could have widespread application, but it is necessarily limited to taking advantage of the components that exist within the BoNT parent proteins. Given that BoNT proteins have evolved as pathogenic factors for clostridia, it is reasonable to assume that there will be application limitations. For example, there exists a major limitation to using clostridial endopeptidases to cleave SNARE proteins and inhibit vesicle trafficking in non-neuronal cells. SNAP-25 is restricted in its expression to neurons, and the ubiquitously expressed homologue, SNAP-23, is not, in man, a substrate for any of the known serotypes. This means that the SNAP-25-cleaving serotypes, BoNT/A, /C and /E, are ineffective at inhibiting secretion and vesicular trafficking in non-neuronal cells. Understanding the structure–function relationship of BoNTs and their substrates has recently led to advances that have taken the first steps to overcoming such limitations, with the reported mutation of a BoNT/E LC so it can cleave human SNAP-23 [9]. Such advancement clearly benefitted from many years of structural biology expertise and progress in the field. In this particular case, the precise understanding of the intra-and intermolecular interactions necessary for the cleavage of SNAP-25 by LC/E was critical to model-led hypotheses that could be tested by wet-laboratory protein engineering. In this example of extending the substrate specificity of BoNT, Chen et al. were able to demonstrate the inhibition of interleukin-8 (IL-8) and mucin release from an appropriately stimulated HeLa cell. The potential to deliver a mutated LC of serotype E and impact SNAP-23-mediated vesicle trafficking is therefore a real possibility and it broadens the spectrum of utility of the BoNTs. Furthermore, using a detailed mutagenesis approach, Wang and colleagues [31] were able to engineer a modified BoNT/C$_1$ LC that was unable to cleave SNAP-25, whilst maintaining syntaxin cleavage ability. Such a molecule has the potential to be used as a tool to dissect SNARE involvement in intracellular biology and, possibly, also to extend the therapeutic potential of BoNTs into disease areas where syntaxin plays a major role.

The combination of structural analysis at the primary, secondary and tertiary level has also elucidated peptide signals within the BoNT sequence that have key biological significance. A good example of such a discovery is provided by studies on di-leucine motifs and ubiquitination signals to engineer duration of action properties into novel proteins. Indeed, the biological advantages of the hybrid molecules reported in the literature [33] are in a major part proposed to be due to the transfer of such motifs within the various BoNT domains. The possibility therefore exists to create bespoke proteins that utilise domains of BoNTs that have been manipulated at the gene level to include domain-or site-specific mutations that afford the novel proteins with new biological properties, for example, in terms of substrate target, duration of action and cellular target.

7.4   Using the Endopeptidase Domain as a Warhead

Understanding the modular structure of the BoNT family, and appreciating the discrete organisation of the LC, H$_N$ and H$_C$ domains, has facilitated a number of engineering approaches that aim to utilise the domains in novel molecule construction. A good example of this approach is provided in the design of new therapeutics based on a BoNT fragment comprising the LC and H$_N$ domains termed LH$_N$. Variously referred to as TSIs or targeted vesicular exocytosis-modulating protein (TVEMP), this novel approach to therapeutic development harnesses the power of the endopeptidase domain to modify the intracellular processes of the target cells and inhibit secretion. This approach therefore aims to extend the therapeutic application of BoNT-derived activities beyond the neuromuscular junction. The manipulation of toxins to create new therapeutic entities has been the subject of much research and development effort

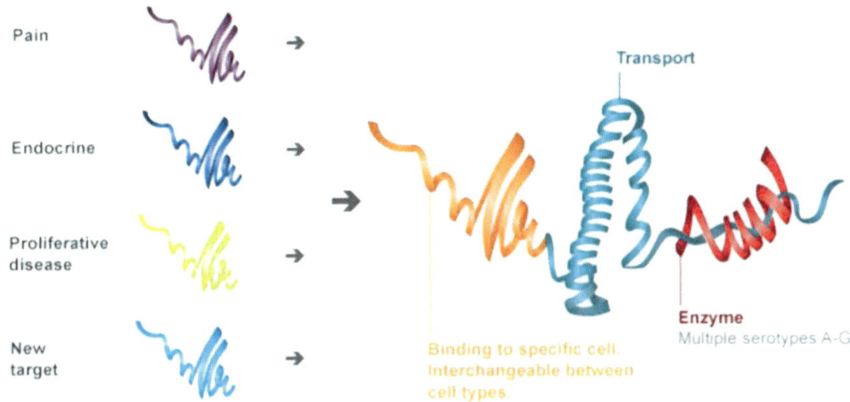

Fig. 7.1 Schematic representation of botulinum neurotoxin structure in relation to engineered targeted secretion inhibitor (TSI) proteins

over many years, most notably in the cancer field with the development of diphtheria toxin-and *Pseudomonas* exotoxin-derived therapeutics. Where the TSI strategy makes advances on current targeted cell ablation-style therapeutics is that the botulinum LC mechanism of action leads to cell manipulation (inhibition of secretion and membrane receptor/channel presentation) rather than cell death. As described previously, BoNTs have evolved an exquisite selectivity towards neuronal cells, by virtue of their selective binding and their neuronally focused substrate cleavage specificities. Delivering the endopeptidase domain to non-neuronal cells not targeted naturally by BoNTs is one way of generating more widely applicable therapeutics. By creating novel recombinant proteins, it has proved possible to deliver the LC into cell types not sensitive to native BoNTs, cleave the relevant SNARE protein and thereby inhibit secretion from a range of otherwise resistant cells. This has enhanced the utilisation of the BoNT domains far beyond that achievable by natural selection.

TSIs are a novel class of multidomain proteins that comprise three basic domains, each providing a contribution to the function of the whole molecule [15]. The overall architecture of the TSI platform is illustrated in Fig. 7.1. Firstly, there is the LC domain of one of the BoNT serotypes, providing the TSI with a SNARE cleavage capability that is dependent on the LC chosen. Secondly, there is the H_N domain which provides the intracellular translocation ability for the LC. Thirdly, there is a binding domain which could be derived from BoNT but is more commonly specified from a peptide or a protein that interacts with a receptor of choice on the target cell. The first report demonstrating that a BoNT LC endopeptidase could be delivered into a target cell via a non-native binding ligand used a chemical conjugate of the nerve growth factor (NGF) and the LH_N/A fragment of BoNT/A to cleave SNAP-25 and inhibit noradrenaline release from PC12 cells [5]. This work established that retargeting the BoNT LH$_N$ fragment was a possibility and, importantly, that the LC could cleave intracellular SNARE proteins in a manner that was comparable to the situation that exists with BoNT intoxication. Subsequently, a conjugate of wheat germ agglutinin (WGA) and the LH$_N$/A fragment was constructed and was observed to deliver the endopeptidase into neuronal and non-neuronal cell types with a consequent cleavage of SNAP-25 and inhibition of secretion [6]. In the latter case, in which HIT-T15 cells were used (naturally resistant to the effects of BoNT/A), this result demonstrated that it is possible to internalise the endopeptidase into the cytosol of a cell normally resistant to the effect of BoNT. This confirmed the ability of the H$_N$ domain to function in the new target cell following binding and endocytosis and truly demonstrated the scope of TSI beyond the neuron.

With proof of principle established for the retargeting approach by NGF and WGA conjugates, a therapeutically relevant application of this approach was developed, targeting a conjugate of *Erythrina crista-galli* lectin and LH$_N$/A (ECL–LH$_N$/A) to nociceptive afferents with the intention of inhibiting the release of neurotransmitters from the nociceptors and consequently inhibition of pain. The properties of ECL–LH$_N$/A were explored in a range of in vitro model systems, and inhibition of both substance P and glutamate release from cultured embryonic dorsal root ganglion neurons was observed for at least 25 days following a single treatment. Intrathecally administered ECL–LH$_N$/A significantly reduced the nociceptive inputs to convergent dorsal horn neurons by primary sensory afferents of the C-fibre and Aδ types, whereas there was little or no effect on sensory inputs from Aβ-fibres [12]. Intrathecal ECL–LH$_N$/A also resulted in prolonged withdrawal latency in a 'hotplate' model of acute thermal pain. This effect was sustained for more than 30 days post administration of the conjugate [8].

These studies with lectin–LH$_N$ conjugates further demonstrated the potential of the retargeting technology and established TSI as a platform with real potential to create novel medicines for a range of conditions, though at this stage the exemplification was primarily through neuronal cell targets. The potential to create novel proteins that enable delivery of a BoNT LC to a diverse range of cell types is now well established. The ability of such proteins to produce pharmacological effects in disease-relevant animal models has also established the therapeutic potential of this approach. The various studies have demonstrated that the retargeted endopeptidase proteins retain the prolonged duration of action that is the hallmark of BoNTs. This means that recombinant proteins based upon this approach are particularly suitable for treating chronic diseases. Chemical conjugates of LH$_N$ fragments and protein ligands enabled the creation of hybrid proteins to deliver a clostridial endopeptidase into specific target cells. However, there are several drawbacks to using chemical conjugation to produce therapeutic proteins, a major one being the inevitable heterogeneous mixture of species that is created using this process and the difficulties of controlling such a process within the manufacturing environment. To progress such a strategy of retargeted BoNT fragments, it was necessary to develop a recombinant platform that would underpin the TSI approach.

The recombinant expression of a catalytically active, stable LH$_N$ fragment of BoNT/A was first reported in 2002 [7]. Subsequently, the expression and purification of LH$_N$/B and LH$_N$/C from *Escherichia coli* was also reported [29]. In all cases, the

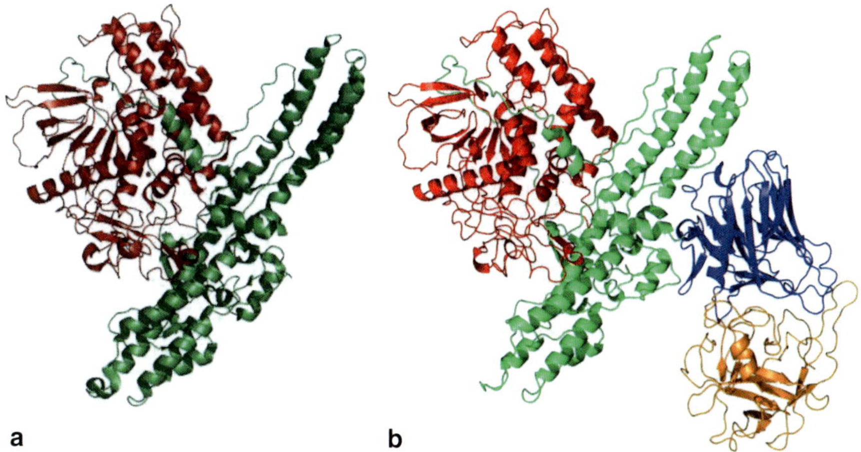

Fig. 7.2 Tertiary structure of the LH$_N$ fragment of botulinum toxin of serotype A (BoNT/A) (**a**) and BoNT/A holotoxin (**b**). The light chain (LC) is depicted in *red*, H$_N$ domain in *green*, H$_{CN}$ in *blue* and H$_{CC}$ in *orange*

recombinant LH$_N$ fragments had very low toxicity because these lack the necessary H$_C$ domain with which to bind to acceptors on the neuronal surface. Despite the lack of an H$_C$ domain, such recombinant LH$_N$ proteins were demonstrated to be stable, catalytically active and effective at intracellular cleavage of the target SNAREs. The relative stability and similarity to the parent neurotoxins are supported by recent studies that have shown recombinant LH$_N$/A [21] and LH$_N$/B [22] to retain the structure of the equivalent domains in the intact BoNT protein. Even though the LH$_N$ fragment lacks the $\sim$50-kDa H$_C$ binding domain and inter-domain interactions therein, this surprising observation highlights the stability of such a molecule and therefore the benefits of such a molecule as a backbone for therapeutic molecule development. The common structural arrangement of the LH$_N$ and BoNT also provides guidance to site-specific LH$_N$ mutagenesis strategies based on BoNT knowledge. Figure 7.2 illustrates the significant structural similarity between LH$_N$/A and BoNT/A.

Given the size and complexity of such a fusion protein, developing a fully recombinant chimera protein, which incorporated the translocation and endopeptidase domains of a BoNT combined with a peptide-targeting ligand, was a challenging task. Nevertheless, a fully recombinant fusion protein consisting of the LH$_N$ fragment of BoNT/C$_1$ and epidermal growth factor (EGF) has been reported [16]. Such a molecule combined an ability to bind EGF receptors with the syntaxin-cleavage activity of the BoNT C LC, thereby providing a novel molecule for the inhibition of secretory events from cells expressing EGF receptors. Creating fully recombinant proteins that target and deliver LC into a specified cell represents a tremendous opportunity to develop therapeutic proteins that inhibit secretion from cells involved in a wide variety of diseases.

Developing the TSI platform further, Syntaxin Ltd have reported that such a platform can lead to a portfolio of engineered molecules that have the potential to lead to novel biopharmaceuticals for use in a wide range of diseases. Harnessing the LC endopeptidase and H_N translocation activities, a series of innovative proteins have been created that are reported to have potential in the treatment of pain, endocrine disease (acromegaly) and cancer. As reported by Somm and colleagues [28], recombinant TSI proteins have been created that specifically target the somatotroph cells of the pituitary. Such molecules have been designed to inhibit the excessive release of growth hormone from pituitary adenomas and are therefore intended for development to treat conditions such as acromegaly. These studies report that engineered TSIs do indeed decrease the circulating levels of growth hormone for many days when assessed in various preclinical studies. Interestingly, such effects are achieved only after a single bolus of test material (rather than continuous infusion), and the material was administered systemically (intravenous; i.v.) rather than being limited to local administration at the site of therapeutic benefit. Such a novel administration route for a molecule derived from BoNT indicates the significant progress that has been made to (1) engineer out unwanted toxic effects that would be inherent in BoNT and (2) engineer in a high level of cell specificity whilst (3) retaining the powerful biochemical mechanisms of substrate cleavage and membrane translocation.

The development of the TSI platform from a concept to a preclinical reality in situations where BoNT would be clinically ineffective is a significant step forward. Though details of targeting domains are not always available, it is noted that natural peptides and proteins are obvious candidates for achieving effective receptor–ligand interactions and so ensuring that the TSI locates the target cell. Natural ligands can have disadvantages or may not be suitable for recombinant expression; however, the growing field of protein scaffolds and recombinant antibody technologies are possible routes to expanding the breadth of targeting domains. It is almost inevitable that the field will take such a direction in order to ensure target cell specificity is achieved. Many of these scaffolds have pre-existing tertiary structural data available and their binding interfaces have been extensively studied. The opportunity therefore exists to combine the BoNT and scaffold protein structural information and facilitate the optimal engineering of the products.

7.5 Clinical Status of Retargeted Endopeptidases

As described in Chaps. 3–6 of this volume, the medical applications of BoNTs are expanding widely, and the number of approved indications is slowly increasing. Nevertheless, the range of indications suitable for BoNT or engineered BoNT therapies will likely be restricted to those with a neuronal basis unless efficient engineering of the H_C domain can be used to establish a BoNT–scaffold platform. The previous section discusses the opportunities that exist with the TSI platform and notes the significant advancement that has been made preclinically. But are there clinical data to support the BoNT fragment-retargeting approach, in the way that fragments of diphtheria and *Pseudomonas* toxins have been used to develop novel targeted cell ablation products? It is exciting to note that clinical studies are under way.

In the first quarter of 2011, Syntaxin Ltd announced that its partner Allergan had initiated two phase II trials to evaluate the safety and efficacy of its retargeted endopeptidase drug candidate AGN- 214868. The phase II trials are focused on patients with post-herpetic neuralgia (PHN; ClinicalTrials.gov Identifier NCT01129531) and idiopathic overactive bladder and urinary incontinence (ClinicalTrials.gov Identifier NCT01157377). With the initiation of phase II trials, the TSI technology platform reached a significant point of development since the AGN- 214868 candidate was discovered under the collaboration using Syntaxin's proprietary discovery platform. Having successfully achieved a first period of data acquisition, the PHN phase II trial has been extended to a second period and is estimated to complete in August 2015. Meanwhile, the overactive bladder phase II studies are estimated to complete late 2013.

7.6 Using the Binding Domain as a Delivery Vehicle

Given that BoNTs very effectively deliver a biologically active effector protein to a selected cell population, the defined neuronal target cells, there has been a significant interest in the potential of the neurotoxins to deliver therapeutic molecules, particularly biological molecules, into nerve cells. BoNTs could deliver to peripheral neurones, particularly cholinergic terminals through the H_C fragment of the neurotoxin. The H_C fragment would be sufficient to target the nerve cell but, without the translocation domain, it would not deliver to the cytosol. Even full-length HCs may be insufficient to achieve cytosolic delivery, presumably reflecting the close co-operative nature of the interaction of the LC and H_N domains during translocation. Further, the most recent understanding of the properties of the H_N domain and its requirements for redox, pH and electrical potential differentials indicate the importance of the H_N domain and particularly the belt component. For these reasons, full-length BoNTs lacking a functional endopeptidase have been proposed as preferred delivery tools [1], [33].

The first report of the cytoplasmic delivery of a cargo protein using BoNT HC was by Weller et al. [34]. These authors reported the functional delivery of tetanus neurotoxin (TeNT) LC into the phrenic nerve by disulphide attachment to the HC of BoNT/A. This demonstrated that the BoNT holotoxin was capable of translocating a non-BoNT, albeit related, protein. One application of BoNTs for the delivery of biomolecules to neurons that has been studied is the replacement of missing enzyme activities, for example, in lysosomal storage diseases and treatment of oxidative injury. Bade et al. [1] demonstrated the ability of BoNT/D to deliver a range of cargo proteins to neurones and achieve enzymatic activity in the neuronal cytosol. Unfolding of the cargo protein was necessary for translocation into the nerve cell, and cargo proteins that were insufficiently flexible in their conformation were not well transported. Understanding the tertiary structure and domain organisation of BoNTs has no doubt assisted with the interpretation of the success of such complex multidomain delivery vehicles, although further understanding is required of the translocation domain and its impact on facilitating the passage of non-clostridial proteins.

A variety of delivery approaches have been summarised by Pickett [23] in which BoNTs have been developed that employ the cellular binding activity of BoNT as a targeting moiety to deliver the activity of a conjugated, non-native protein. As biochemical and structural studies have shown that TeNT has a similar modular arrangement of functional domains as BoNT, Pickett notes that advances in TeNT engineering may also be applied to BoNT. For example, the binding domains of both BoNT and TeNT have been utilised to deliver DNA to target cells and enhance the targeting of other transfection methods. The TeNT HC fragment has been conjugated to polylysine, which has a high capacity to bind DNA, allowing the transfection of a range of neuronal cell lines [2]. Recent studies with recombinantly produced BoNT domains show that proteins can be assembled by non-chemical linking, using tagging with helical motifs from the family of SNARE proteins. Again, structural input into the relationship between primary sequence and function has been instrumental in understanding the potential for SNARE-dependent coupling of proteins. Such a strategy may potentially be exploited to use the BoNT-binding domain to deliver future therapeutics or other cargo into neurons [10]. Indeed, Ferrari and colleagues have recently reported in vivo studies that explored the functionality of a BoNT that had been reconstituted using the SNARE protein-based 'protein-stapling' technology [14].

In a further application of the BoNT-binding domain as a delivery vehicle, Oyler and colleagues [19] have postulated the use of BoNT or BoNT HC fragments to deliver 'targeted F-box' agents to BoNT-intoxicated neurons as a potential therapeutic for BoNT poisoning. For many years, the logic of using the BoNT-binding domain to deliver BoNT-poisoning 'rescue' molecules has been the subject of significant interest. The F-box strategy is hypothesised to be successful because the targeted F-box agents cause increased ubiquitination and accelerate the turnover of the targeted BoNT/A protease within neurons. Although elegant in design, the strategy critically requires targeted delivery to the correct neuronal population, and BoNT (or fragments of BoNT) would seem to continue to be the best candidate to achieve this.

In summary, various cargos have been proposed for delivery by suitable non-toxic BoNT-based vehicles for the treatment of diseased neuromuscular junctions or to enhance motor neuron function. These include anti-neurotoxin therapies, anti-neurotrophic viral treatments, neuronal enzyme replacement, ion channel modulators, neurotrophic factor receptor modulators and protein replacement for hereditary or autoimmune presynaptic disorders. Whilst an attractive opportunity supported by model system studies, there are not yet any clinical applications of drug delivery using a BoNT-based vehicle and more needs to be understood in structural terms for the translocation domain and how this impacts on its limitations.

7.7 Conclusions

It is clear that the field of BoNT engineering has progressed significantly in the past few years and it is starting to realise the potential for the development of new products with novel biological mechanisms. It is also well understood that BoNT/A is a major therapeutic product that can be widely used to treat various neurological and

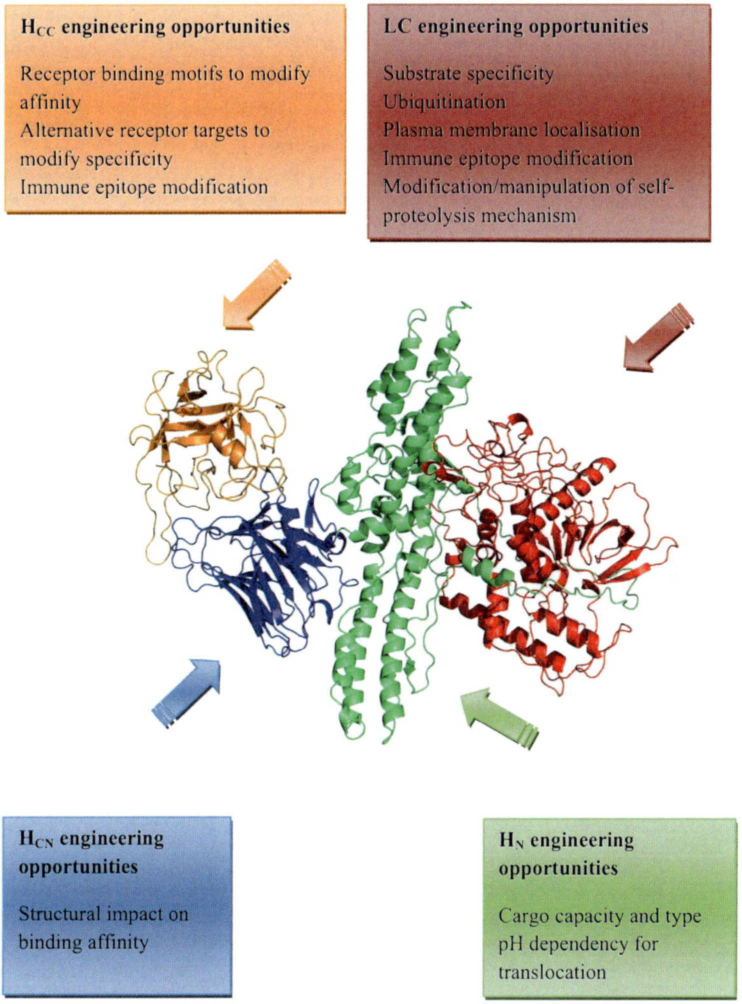

H_{CC} engineering opportunities

Receptor binding motifs to modify affinity
Alternative receptor targets to modify specificity
Immune epitope modification

LC engineering opportunities

Substrate specificity
Ubiquitination
Plasma membrane localisation
Immune epitope modification
Modification/manipulation of self-proteolysis mechanism

H_{CN} engineering opportunities

Structural impact on binding affinity

H_N engineering opportunities

Cargo capacity and type
pH dependency for translocation

Fig. 7.3 The potential for engineering botulinum neurotoxin (BoNT) domains

neuromuscular conditions. The therapeutic success of the BoNTs results from their specific and potent inhibition of neurotransmitter release from peripheral cholinergic neurons combined with a duration of action measured in months. The clinical utility of the neurotoxins is, however, severely constrained, by both their limited range of target cells and narrow therapeutic window. Advances over the past 20 years in understanding the structure and biology of BoNTs, combined with the developments in recombinant protein engineering, are opening up opportunities to engineer novel therapeutic proteins based upon the unique pharmacological properties of the BoNTs (Fig. 7.3). One such opportunity, TSI, is already progressing in the clinical setting and preclinically for a range of indications not treatable with neurotoxin products.

Similarly, the progress being made to delineate the LC–substrate inter-relationship is providing innovative proteins with unique modes of action. These developments will increase the medical benefits achieved through the clinical application of the native neurotoxins, particularly BoNT/A, while removing the inherent toxicity and providing proteins with a much-improved therapeutic window. Harnessing the properties of the neurotoxins' protein domains in novel recombinant proteins will lead to the creation of a completely new class of biologics. By inhibiting secretion, these will treat chronic conditions like chronic pain, which currently have few effective treatments. Throughout all of this development activity, understanding the relationship between toxin structure (primary, secondary and tertiary) and function has been integral; without doubt, it will be essential to maximising the potential of this exciting class of proteins.

References

1. Bade S, Rummel A, Reisinger C, Karnath T, Ahnert-Hilger G, Bigalke H, Binz T (2004) Botulinum neurotoxin type D enables cytosolic delivery of enzymatically active cargo proteins to neurones via unfolded translocation intermediates. J Neurochem 91(6):1461–1472
2. Box M, Parks DA, Knight A, Hale C, Fishman PS, Fairweather NF (2003) A multi-domain protein system based on the HC fragment of tetanus toxin for targeting DNA to neuronal cells. J Drug Target 11(6):333–343
3. Carter AT, Paul CJ, Mason DR, Twine SM, Alston MJ, Logan SM, Austin JW, Peck MW (2009) Independent evolution of neurotoxin and flagellar genetic loci in proteolytic Clostridium botulinum. BMC Genomics 10:115
4. Chaddock JA, Acharya KR (2011) Engineering toxins for 21st century therapies. FEBS J 278(6):899–904
5. Chaddock JA, Purkiss JR, Duggan MJ, Quinn CP, Shone CC, Foster KA (2000a) A conjugate composed of nerve growth factor coupled to a non-toxic derivative of *Clostridium botulinum* neurotoxin type A can inhibit neurotransmitter release in vitro. Growth Factors 18(2):147–155
6. Chaddock JA, Purkiss JR, Friis LM, Broadbridge JD, Duggan MJ, Fooks SJ, Shone CC, Quinn CP, Foster KA (2000b) Inhibition of vesicular secretion in both neuronal and nonneuronal cells by a retargeted endopeptidase derivative of *Clostridium botulinum* neurotoxin type A. Infect Immun 68(5):2587–2593
7. Chaddock JA, Herbert MH, Ling RJ, Alexander FCG, Fooks SJ, Revell DF, Quinn CP, Shone CC, Foster KA (2002) Expression and purification of catalytically active, non-toxic endopeptidase derivatives of Clostridium botulinum toxin type A. Protein Expr Purif 25:219–228
8. Chaddock JA, Purkiss JR, Alexander FC, Doward S, Fooks SJ, Friis LM, Hall YH, Kirby ER, Leeds N, Moulsdale HJ, Dickenson A, Green GM, Rahman W, Suzuki R, Duggan MJ, Quinn CP, Shone CC, Foster KA (2004) Retargeted clostridial endopeptidases: inhibition of nociceptive neurotransmitter release in vitro, and antinociceptive activity in in vivo models of pain. Mov Disord 19(Suppl 8):42–47
9. Chen S, Barbieri JT (2009) Engineering botulinum neurotoxin to extend therapeutic intervention. Proc Natl Acad Sci U S A 106(23):9180–9184
10. Darios F, Niranjan D, Ferrari E, Zhang F, Soloviev M, Rummel A, Bigalke H, Suckling J, Ushkaryov Y, Naumenko N, Shakirzyanova A, Giniatullin R, Maywood E, Hastings M, Binz T, Davletov B (2010) SNARE tagging allows stepwise assembly of a multimodular medicinal toxin. Proc Natl Acad Sci U S A 107(42):18197–18201

11. Dolly JO, Wang J, Zurawski TH, Meng J (2011) Novel therapeutics based on recombinant botulinum neurotoxins to normalize the release of transmitters and pain mediators. FEBS J 278(23):4454–4466
12. Duggan MJ, Quinn CP, Chaddock JA, Purkiss JR, Alexander FC, Doward S, Fooks SJ, Friis LM, Hall YH, Kirby ER, Leeds N, Moulsdale HJ, Dickenson A, Green GM, Rahman W, Suzuki R, Shone CC, Foster KA (2002) Inhibition of release of neurotransmitters from rat dorsal root ganglia by a novel conjugate of a *Clostridium botulinum* toxin A endopeptidase fragment and Erythrina cristagalli lectin. J Biol Chem 277(38):34846–34852
13. Erbguth FJ (2004) Historical notes on botulism, *Clostridium botulinum*, botulinum toxin, and the idea of the therapeutic use of the toxin. Mov Disord 19(Suppl 8):2–6
14. Ferrari E, Maywood ES, Restani L, Caleo M, Pirazzini M, Rossetto O, Hastings MH, Niranjan D, Schiavo G, Davletov B (2011) Re-assembled botulinum neurotoxin inhibits CNS functions without systemic toxicity. Toxins 3(4):345–355
15. Foster KA, Chaddock JA (2010) Targeted secretion inhibitors—innovative protein therapeutics. Toxins 2(12):2795–2815
16. Foster KA, Adams EJ, Durose L, Cruttwell CJ, Marks E, Shone CC, Chaddock JA, Cox CL, Heaton C, Sutton JM, Wayne J, Alexander FC, Rogers DF (2006) Re-engineering the target specificity of Clostridial neurotoxins—a route to novel therapeutics. Neurotox Res 9(2–3): 101–107
17. Henkel JS, Jacobson M, Tepp W, Pier C, Johnson Ea, Barbieri JT (2009) Catalytic properties of botulinum neurotoxin subtypes A3 and A4. Biochemistry 48:2522–2528
18. Kumaran D, Eswaramoorthy S, Furey W, Navaza J, Sax M, Swaminathan S (2009) Domain organization in *Clostridium botulinum* neurotoxin type E is unique: its implication in faster translocation. J Mol Biol 386:233–245
19. Kuo CL, Oyler GA, Shoemaker CB (2011) Accelerated neuronal cell recovery from botulinum neurotoxin intoxication by targeted ubiquitination. PLoS One 6(5):e20352
20. Lacy DB, Tepp W, Cohen AC, DasGupta BR, Stevens RC (1998) Crystal structure of botulinum neurotoxin type A and implications for toxicity. Nat Struct Biol 5(10):898–902
21. Masuyer G, Thiyagarajan N, James PL, Marks PM, Chaddock JA, Acharya KR (2009) Crystal structure of a catalytically active, non-toxic endopeptidase derivative of *Clostridium botulinum* toxin A. Biochem Biophys Res Commun 381(1):50–53
22. Masuyer G, Beard M, Cadd VA, Chaddock JA, Acharya KR (2011) Structure and activity of a functional derivative of *Clostridium botulinum* neurotoxin B. J Struct Biol 174(1):52–57
23. Pickett A, Perrow K (2011) Towards new uses of botulinum toxin as a novel therapeutic tool. Toxins 3:63–81
24. Pier CL, Chen C, Tepp WH, Lin G, Janda KD, Barbieri JT, Pellett S, Johnson EA (2011) Botulinum neurotoxin subtype A2 enters neuronal cells faster than subtype A1. FEBS Lett 585(1):199–206
25. Rummel A, Mahrhold S, Bigalke H, Binz T (2004) The HCC-domain of botulinum neurotoxins A and B exhibits a singular ganglioside binding site displaying serotype specific carbohydrate interaction. Mol Microbiol 51(3):631–643
26. Rummel A, Mahrhold S, Bigalke H, Binz T (2011) Exchange of the H(CC) domain mediating the double receptor recognition improves the pharmacodynamic properties of botulinum neurotoxin. FEBS J 278(23):4506–4515
27. Scott AB (1980) Botulinum toxin injection into extraocular muscles as an alternative to strabismus surgery. J Pediatr Ophthalmol Strabismus 17(1):21–25
28. Somm E, Bonnet N, Martinez A, Marks PM, Cadd VA, Elliott M, Toulotte A, Ferrari SL, Rizzoli R, Hüppi PS, Harper E, Melmed S, Jones R, Aubert ML (2012) A botulinum toxin-derived targeted secretion inhibitor downregulates the GH/IGF1 axis. J Clin Invest 122(9):3295–3306
29. Sutton JM, Wayne J, Scott-tucker A, Brien SMO, Marks PMH, Alexander FCG, Shone CC, Chaddock JA (2005) Preparation of specifcally activatable endopeptidase derivatives of *Clostridium botulinum* toxins type A, B, and C and their applications. 40:31–41
30. Swaminathan S, Eswaramoorthy S (2000) Crystallization and preliminary X-ray analysis of *Clostridium botulinum* neurotoxin type B. Acta Crystallogr D Biol Crystallogr 56(Pt 8): 1024–1026

31. Wang D, Zhang Z, Dong M, Sun S, Chapman ER, Jackson MB (2011) Syntaxin requirement for Ca^{2+}-triggered exocytosis in neurons and endocrine cells demonstrated with an engineered neurotoxin. Biochemistry 50(14):2711–2713
32. Wang J, Meng J, Lawrence GW, Zurawski TH, Sasse A, Bodeker MO, Gilmore MA, Fernandez-Salas E, Francis J, Steward LE, Aoki KR, Dolly JO (2008) Novel chimeras of botulinum neurotoxins A and E unveil contributions from the binding, translocation, and protease domains to their functional characteristics. J Biol Chem 283(25):16993–17002
33. Wang J, Zurawski TH, Meng J, Lawrence G, Olango WM, Finn DP, Wheeler L, Dolly JO (2011) A dileucine in the protease of botulinum toxin A underlies its long-lived neuroparalysis: transfer of longevity to a novel potential therapeutic. J Biol Chem 286(8):6375–6385
34. Weller U, Dauzenroth ME, Gansel M, Dreyer F (1991) Cooperative action of the light chain of tetanus toxin and the heavy chain of botulinum toxin type A on the transmitter release of mammalian motor endplates. Neurosci Lett 122(1):132–134

Index

K. A. Foster (ed.), *Clinical Applications of Botulinum Neurotoxin,*
Current Topics in Neurotoxicity 5, DOI 10.1007/978-1-4939-0261-3,
© Springer Science+Business Media New York 2014